高等学校专业英语系列教材

建筑环境与设备工程专业

张寅平　潘毅群　王　馨　编著
赵荣义　审校

中国建筑工业出版社

图书在版编目（CIP）数据

建筑环境与设备工程专业/张寅平等编著.—北京：
中国建筑工业出版社，2004（2024.7重印）
（高等学校专业英语系列教材）
ISBN 978-7-112-06643-8

Ⅰ.建… Ⅱ.张… Ⅲ.建筑工程—英语—高等学校—教材 Ⅳ.TU

中国版本图书馆 CIP 数据核字(2004)第108320号

责任编辑：齐庆梅
责任设计：孙　梅
责任校对：刘玉英　王金珠

高等学校专业英语系列教材
建筑环境与设备工程专业
张寅平　潘毅群　王　馨　编著
赵荣义　审校

*

中国建筑工业出版社出版、发行（北京西郊百万庄）
各地新华书店、建筑书店经销
北京市密东印刷有限公司印刷

*

开本：787×1092 毫米　1/16　印张：17½　字数：435 千字
2005 年 1 月第一版　2024 年 7 月第十七次印刷
定价：**30.00** 元
ISBN 978-7-112-06643-8
(20986)

版权所有　翻印必究
如有印装质量问题，可寄本社退换
(邮政编码　100037)

前言

本书阅读材料主要选自美国暖通空调制冷机械工程师协会(American Society of Heating, Refrigerating, and Air-conditioning，简称 ASHRAE) 1999～2003 年的 4 卷手册(ASHRAE Handbooks)，该手册每年 1 卷，4 年一轮回，内容包括暖通空调制冷领域的基础理论(Fundamentals)、制冷(Refrigeration)与暖通空调的应用(HVAC Application)和暖通空调系统及设备(HVAC Systems and Equipment)。该套手册内容系统、表述清晰、图表精美、语言规范，是暖通空调制冷领域难得的专业英语阅读材料库。

结合几年来本课程的教学实践，编者在选材上作了取舍：注重专业基础内容和暖通空调制冷领域目前的热点科研方向。所列生词、短语和注释的内容也是教学实践中同学经常提出的问题，因此针对性较强。此外，每章还安排了习题。通过上述材料的阅读和练习，读者可对专业英语"窥一斑而知全豹"，为其阅读英语专业论文打下良好基础。全书最后，简单介绍了英文科研论文写作方面的基本知识，旨在对建筑环境与设备工程领域英文科技论文的写作有所帮助。附录中还对建筑环境与设备工程专业国际学术期刊、建筑环境与设备工程专业国际学术组织以及国际学术期刊检索系统做了简要介绍。

全书编写分工如下：王馨(清华大学)编写了 Lesson 1～Lesson 5，潘毅群(同济大学)编写了 Lesson 6～Lesson 12，张寅平(清华大学)负责全书选材并撰写了英语科研论文写作简介、附录。杨瑞、徐秋健为全书提供了大量素材，陈斌娇做了大量编辑工作。

全书每课均有习题并附有参考答案。练习中出现的部分句子和短语，在课文中用下划线标出，一些专业名词也在课文中用黑体表示。课文中出现的"Chapter"与 ASHRAE 原文对应。

清华大学赵荣义教授花费了大量时间，仔细审校了全书，对原稿进行了认真修改。此外，中国建筑工业出版社齐庆梅编辑在本书编写工作中提供了许多帮助。在此，编者表示由衷的感谢。

本书可作为建筑环境与设备工程专业及其相关专业研究生和本科生专业英语阅读和写作课的教材，任课教师可根据具体要求选择所需内容。同时还可供上述专业的教师、科研人员和工程技术人员参考。由于编者水平有限，编写中难免有不妥之处，衷心欢迎广大师生指正。

Contents

Part Ⅰ Theory ·· 1

   LESSON 1   THERMODYNAMICS AND REFRIGERATION
                     CYCLES ··· 3
       WORDS AND EXPRESSIONS ·· 25
       NOTATIONS ·· 26
       EXERCISES ··· 26
   LESSON 2   FLUID FLOW ··· 28
       WORDS AND EXPRESSIONS ·· 42
       NOTATIONS ·· 44
       EXERCISES ··· 45
   LESSON 3   HEAT TRANSFER ·· 47
       WORDS AND EXPRESSIONS ·· 71
       NOTATIONS ·· 73
       EXERCISES ··· 74
   LESSON 4   MASS TRANSFER ·· 76
       WORDS AND EXPRESSIONS ·· 93
       NOTATIONS ·· 94
       EXERCISES ··· 95
   LESSON 5   PSYCHROMETRICS ··· 96
       WORDS AND EXPRESSIONS ·· 109
       NOTATIONS ·· 110
       EXERCISES ··· 111

Part Ⅱ General Engineering Information ·································· 113

   LESSON 6   THERMAL COMFORT ··· 115
       WORDS AND EXPRESSIONS ·· 129
       NOTATIONS ·· 131
       EXERCISES ··· 132
   LESSON 7   AIR CONTAMINANTS ··· 133
       WORDS AND EXPRESSIONS ·· 146

  NOTATIONS ……………………………………… 149
  EXERCISES ……………………………………… 152
Part Ⅲ HVAC systems and equipment ……………… 153
 LESSON 8 CENTRAL COOLING AND HEATING ……… 155
  WORDS AND EXPRESSIONS ………………… 163
  NOTATIONS ……………………………………… 164
  EXERCISES ……………………………………… 166
 LESSON 9 DECENTRALIZED COOLING
       AND HEATING ……………………… 167
  WORDS AND EXPRESSIONS ………………… 179
  NOTATIONS ……………………………………… 180
  EXERCISES ……………………………………… 181
 LESSON 10 AIR-COOLING AND DEHUMIDIFYING
        COILS …………………………………… 182
  WORDS AND EXPRESSIONS ………………… 191
  NOTATIONS ……………………………………… 193
  EXERCISES ……………………………………… 195
 LESSON 11 AIR CLEANERS FOR PARTICULATE
        CONTAMINANTS …………………… 196
  WORDS AND EXPRESSIONS ………………… 208
  NOTATIONS ……………………………………… 210
  EXERCISES ……………………………………… 212
 LESSON 12 UNITARY AIR CONDITIONERS AND
        UNITARY HEAT PUMPS …………… 213
  WORDS AND EXPRESSIONS ………………… 224
  NOTATIONS ……………………………………… 225
  EXERCISES ……………………………………… 226
英文科研论文写作简介 ……………………………………… 227
练习参考答案 ………………………………………………… 252
附录 …………………………………………………………… 259
 附录 A 国际相关组织介绍 ………………………… 259
 附录 B 有关国际会议简介 ………………………… 260
 附录 C 相关领域的一些国际期刊简介 …………… 262
 附录 D EI 和 SCI 检索的简易教程 ………………… 266

Part I

Theory

LESSON 1

THERMODYNAMICS AND REFRIGERATION CYCLES

THERMODYNAMICS is the study of energy, its transformations, and its relation to states of matter. This chapter covers the application of thermodynamics to refrigeration cycles. The first part reviews the first and second laws of thermodynamics and presents methods for calculating thermodynamic properties. The second part addresses compression refrigeration cycles.

THERMODYNAMICS

A **thermodynamic system** is a region in space or a quantity of matter bounded by a closed surface. The surroundings include everything external to the system, and the system is separated from the surroundings by the system boundaries. These boundaries can be movable or fixed, real or imaginary.

The concepts that operate in any thermodynamic system are **entropy** and **energy**. Entropy measures the molecular disorder of a system. The more mixed a system, the greater its entropy; conversely, an orderly or unmixed configuration is one of low entropy. Energy has the capacity for producing an effect and can be categorized into either stored or transient forms as described in the following sections.

Stored Energy

Thermal (internal) energy is the energy possessed by a system caused by the motion of the molecules and/or intermolecular forces.

Potential energy is the energy possessed by a system caused by the attractive forces existing between molecules, or the elevation of the system.

$$\text{PE} = mgz \qquad (1)$$

where
   m = mass
   g = local acceleration of gravity
   z = elevation above horizontal reference plane

Kinetic energy is the energy possessed by a system caused by the velocity of the molecules and is expressed as

---

Extracted from Chapter I of the ASHRAE *Handbook-Fundamentals*.

$$KE = mV^2/2 \tag{2}$$

where V is the velocity of a fluid stream crossing the system boundary.

Chemical energy is energy possessed by the system caused by the arrangement of atoms composing the molecules.

Nuclear (atomic) energy is energy possessed by the system from the cohesive forces holding protons and neutrons together as the atom's nucleus.

Energy in Transition

Heat (Q) is the mechanism that transfers energy across the boundary of systems with differing temperatures, always toward the lower temperature. Heat is positive when energy is added to the system(see Figure 1).

Work is the mechanism that transfers energy across the boundary of systems with differing pressures (or force of any kind), always to-

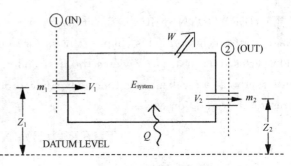

Fig. 1 Energy Flows in General Thermodynamic System

ward the lower pressure. If the total effect produced in the system can be reduced to the raising of a weight, then nothing but work has crossed the boundary. Work is positive when energy is removed from the system (see Figure 1).

Mechanical or **shaft work (W)** is the energy delivered or absorbed by a mechanism, such as a turbine, air compressor, or internal combustion engine.

Flow work is energy carried into or transmitted across the system boundary because a pumping process occurs somewhere outside the system, causing fluid to enter the system. It can be more easily understood as the work done by the fluid just outside the system on the adjacent fluid entering the system to force or push it into the system. Flow work also occurs as fluid leaves the system.

$$\text{Flow Work(per unit mass)} = pv \tag{3}$$

where p is the pressure and v is the specific volume, or the volume displaced per unit mass.

A **property** of a system is any observable characteristic of the system. The **state** of a system is defined by listing its properties. The most common thermodynamic properties are temperature T, pressure p, and specific volume v or density ρ. Additional thermodynamic properties include entropy, stored forms of energy, and enthalpy.

Frequently, thermodynamic properties combine to form other properties. **Enthalpy (h)**, a result of combining properties, is defined as

$$h \equiv u + pv \tag{4}$$

where u is internal energy per unit mass.

Each property in a given state has only one definite value, and any property always has the same value for a given state, regardless of how the substance arrived at that state.

A **process** is a change in state that can be defined as any change in the properties of a system. A process is described by specifying the initial and final equilibrium states, the path (if identifiable), and the interactions that take place across system boundaries during the process.

A **cycle** is a process or a series of processes wherein the initial and final states of the system are identical. Therefore, at the conclusion of a cycle, all the properties have the same value they had at the beginning.

A **pure substance** has a homogeneous and invariable chemical composition. It can exist in more than one phase, but the chemical composition is the same in all phases.

If a substance exists as liquid at the saturation temperature and pressure, it is called **saturated liquid.** If the temperature of the liquid is lower than the saturation temperature for the existing pressure, it is called either a **subcooled liquid** (the temperature is lower than the saturation temperature for the given pressure) or a **compressed liquid** (the pressure is greater than the saturation pressure for the given temperature).

When a substance exists as part liquid and part vapor at the saturation temperature, its quality is defined as the ratio of the mass of vapor to the total mass. Quality has meaning only when the substance is in a saturated state; i. e., at saturation pressure and temperature.

If a substance exists as vapor at the saturation temperature, it is called **saturated vapor.** (Sometimes the term **dry saturated vapor** is used to emphasize that the quality is 100%.) When the vapor is at a temperature greater than the saturation temperature, it is **superheated vapor.** The pressure and temperature of superheated vapor are independent properties, since the temperature can increase while the pressure remains constant. Gases are highly superheated vapors.

FIRST LAW OF THERMODYNAMICS

The first law of thermodynamics is often called the **law of the conservation of energy.** The following form of the first law equation is valid only in the absence of a nuclear or chemical reaction.

Based on the first law or the law of conservation of energy for any system, open or closed, there is an energy balance as

$$\begin{bmatrix} \text{Net Amount of Energy} \\ \text{Added to Sysem} \end{bmatrix} = \begin{bmatrix} \text{Net Increase in Stored} \\ \text{Energy of System} \end{bmatrix}$$

or

$$\text{Energy In} - \text{Energy Out} = \text{Increase in Energy in Sysem}$$

Figure 1 illustrates energy flows into and out of a thermodynamic system. For the general case of multiple mass flows in and out of the system, the energy balance can be written

$$\Sigma m_{in}\left(u+pv+\frac{V^2}{2}+gz\right)_{in}-\Sigma m_{out}\left(u+pv+\frac{V^2}{2}+gz\right)_{out}+Q-W$$
$$=\left[m_f\left(u+\frac{V^2}{2}+gz\right)_f-m_i\left(u+\frac{V^2}{2}+gz\right)_i\right]_{system} \qquad (5)$$

where subscripts i and f refer to the initial and final states, respectively.

The steady-flow process is important in engineering applications. Steady flow signifies that all quantities associated with the system do not vary with time. Consequently,

$$\sum_{\substack{\text{all streams}\\\text{leaving}}}\dot{m}\left(h+\frac{V^2}{2}+gz\right)-\sum_{\substack{\text{all streams}\\\text{entering}}}\dot{m}\left(h+\frac{V^2}{2}+gz\right)+\dot{Q}-\dot{W}=0 \qquad (6)$$

where $h=u+pv$ is described in Equation (4).

A second common application is the closed stationary system for which the first law equation reduces to

$$Q-W=[m(u_f-u_i)]_{system} \qquad (7)$$

SECOND LAW OF THERMODYNAMICS

The second law of thermodynamics differentiates and quantifies processes that only proceed in a certain direction (irreversible) from those that are reversible. The second law may be described in several ways. One method uses the concept of entropy flow in an open system and the irreversibility associated with the process. The concept of irreversibility provides added insight into the operation of cycles. For example, the larger the irreversibility in a refrigeration cycle operating with a given refrigeration load between two fixed temperature levels, the larger the amount of work required to operate the cycle. Irreversibilities include pressure drops in lines and heat exchangers, heat transfer between fluids of different temperature, and mechanical friction. Reducing total irreversibility in a cycle improves the cycle performance. In the limit of no irreversibilities, a cycle will attain its maximum ideal efficiency.

In an open system, the second law of thermodynamics can be described in terms of entropy as

$$dS_{system}=\frac{\delta Q}{T}+\delta m_i s_i-\delta m_e s_e+dI \qquad (8)$$

where

    dS_{system} = total change within system in time dt during process

    $\delta m_i s_i$ = entropy increase caused by mass entering (incoming)

    $\delta m_e s_e$ = entropy decrease caused by mass leaving (exiting)

    $\delta Q/T$ = entropy change caused by reversible heat transfer between system and surroundings

dI = entropy caused by irreversibilities (always positive)

Equation (8) accounts for all entropy changes in the system. Rearranged, this equation becomes

$$\delta Q = T\,[(\delta m_e s_e - \delta m_i s_i) + dS_{sys} - dI] \tag{9}$$

In integrated form, if inlet and outlet properties, mass flow, and interactions with the surroundings do not vary with time, the general equation for the second law is

$$(S_f - S_i)_{system} = \int_{rev} \frac{\delta Q}{T} + \Sigma(ms)_{in} - \Sigma(ms)_{out} + I \tag{10}$$

In many applications the process can be considered to be operating steadily with no change in time. The change in entropy of the system is therefore zero. The irreversibility rate, which is the rate of entropy production caused by irreversibilities in the process, can be determined by rearranging Equation (10)

$$\dot{I} = \Sigma(\dot{m}s)_{out} - \Sigma(\dot{m}s)_{in} - \int \frac{\dot{Q}}{T_{surr}} \tag{11}$$

Equation (6) can be used to replace the heat transfer quantity. Note that the absolute temperature of the surroundings with which the system is exchanging heat is used in the last term. If the temperature of the surroundings is equal to the temperature of the system, the heat is transferred reversibly and Equation (11) becomes equal to zero.

Equation (11) is commonly applied to a system with one mass flow in, the same mass flow out, no work, and negligible kinetic or potential energy flows. Combining Equations (6) and (11) yields

$$\dot{I} = \dot{m}\left[(s_{out} - s_{in}) - \frac{h_{out} - h_{in}}{T_{surr}}\right] \tag{12}$$

In a cycle, the reduction of work produced by a power cycle or the increase in work required by a refrigeration cycle is equal to the absolute ambient temperature multiplied by the sum of the irreversibilities in all the processes in the cycle. Thus the difference in the reversible work and the actual work for any refrigeration cycle, theoretical or real, operating under the same conditions becomes

$$\dot{W}_{actual} = \dot{W}_{reversible} + T_0 \Sigma \dot{I} \tag{13}$$

THERMODYNAMIC ANALYSIS OF REFRIGERATION CYCLES

Refrigeration cycles transfer thermal energy from a region of low temperature T_R to one of higher temperature. Usually the higher temperature heat sink is the ambient air or cooling water. This temperature is designated as T_0, the temperature of the surroundings.

The first and second laws of thermodynamics can be applied to individual components to determine mass and energy balances and the irreversibility of the components. This procedure is illustrated in later sections in this chapter.

Performance of a refrigeration cycle is usually described by a **coefficient of perform-**

ance. COP is defined as the benefit of the cycle (amount of heat removed) divided by the required energy input to operate the cycle, or

$$\text{COP} \equiv \frac{\text{Useful refrigerating effect}}{\text{Net energy supplied from external sources}} \tag{14}$$

For a mechanical vapor compression system, the net energy supplied is usually in the form of work, mechanical or electrical, and may include work to the compressor and fans or pumps. Thus

$$\text{COP} = \frac{Q_{\text{evap}}}{W_{\text{net}}} \tag{15}$$

In an absorption refrigeration cycle, the net energy supplied is usually in the form of heat into the generator and work into the pumps and fans, or

$$\text{COP} = \frac{Q_{\text{evap}}}{Q_{\text{gen}} + W_{\text{net}}} \tag{16}$$

In many cases the work supplied to an absorption system is very small compared to the amount of heat supplied to the generator, so the work term is often neglected.

Application of the second law to an entire refrigeration cycle shows that a completely reversible cycle operating under the same conditions has the maximum possible Coefficient of Performance. A measure of the departure of the actual cycle from an ideal reversible cycle is given by the **refrigerating efficiency**:

$$\eta_R = \frac{\text{COP}}{(\text{COP})_{\text{rev}}} \tag{17}$$

The Carnot cycle usually serves as the ideal reversible refrigeration cycle. For multistage cycles, each stage is described by a reversible cycle.

EQUATIONS OF STATE

The equation of state of a pure substance is a mathematical relation between pressure, specific volume, and temperature. When the system is in thermodynamic equilibrium,

$$f(p, v, T) = 0 \tag{18}$$

The principles of statistical mechanics are used to (1) explore the fundamental properties of matter, (2) predict an equation of state based on the statistical nature of a particular system, or (3) propose a functional form for an equation of state with unknown parameters that are determined by measuring thermodynamic properties of a substance. A fundamental equation with this basis is the **virial equation**. The virial equation is expressed as an expansion in pressure p or in reciprocal values of volume per unit mass v as

$$\frac{pv}{RT} = 1 + B'p + C'p^2 + D'p^3 + \cdots \tag{19}$$

$$\frac{pv}{RT} = 1 + (B/v) + (C/v^2) + (D/v^3) + \cdots \tag{20}$$

where coefficients B', C', D', etc., and B, C, D, etc., are the virial coefficients. B' and B are second virial coefficients; C' and C are third virial coefficients, etc. The virial coefficients are functions of temperature only, and values of the respective coefficients in Equations (19) and (20) are related. For example, $B'=B/RT$ and $C'=(C-B^2)/(RT)^2$.

The ideal gas constant R is defined as

$$R=\lim_{p\to 0}\frac{(pv)_T}{T_{tp}} \tag{21}$$

where $(pv)_T$ is the product of the pressure and the volume along an isotherm, and T_{tp} is the defined temperature of the triple point of water, which is 273.16 K. The current best value of R is 8314.41 J/(kg mole · K).

The quantity pv/RT is also called the **compressibility factor**; i.e., $Z=pv/RT$ or

$$z=1+(B/v)+(C/v^2)+(D/v^3)+\cdots \tag{22}$$

An advantage of the virial form is that statistical mechanics can be used to predict the lower order coefficients and provide physical significance to the virial coefficients. For example, in Equation (22), the term B/v is a function of interactions between two molecules, C/v^2 between three molecules, etc. Since the lower order interactions are common, the contributions of the higher order terms are successively less. Thermodynamicists use the partition or distribution function to determine virial coefficients; however, experimental values of the second and third coefficients are preferred. For dense fluids, many higher order terms are necessary that can neither be satisfactorily predicted from theory nor determined from experimental measurements. <u>In general, a truncated virial expansion of four terms is valid for densities of less than one-half the value at the critical point.</u> For higher densities, additional terms can be used and determined empirically.

Digital computers allow the use of very complex equations of state in calculating p-v-T values, even to high densities. The Benedict-Webb-Rubin (B-W-R) equation of state (Benedict et al. 1940) and the Martin-Hou equation (1955) have had considerable use, but should generally be limited to densities less than the critical value. Strobridge (1962) suggested a modified Benedict-Webb-Rubin relation that gives excellent results at higher densities and can be used for a p-v-T surface that extends into the liquid phase.

The B-W-R equation has been used extensively for hydrocarbons (Cooper and Goldfrank 1967):

$$P = (RT/v)+(B_0 RT-A_0-C_0/T^2)/v^2+(bRT-a)/v^3+ \\ (a\alpha)/v^6+[c(1+\gamma/v^2)e^{(-\gamma/v^2)}]/v^3 T^2 \tag{23}$$

where the constant coefficients are A_0, B_0, C_0, a, b, c, α, γ.

The Martin-Hou equation, developed for fluorinated hydrocarbon properties, has been used to calculate the thermodynamic property tables. The Martin-Hou equation is as follows:

$$p=\frac{RT}{v-b}+\frac{A_2+B_2 T+C_2 e^{(-kT/T_c)}}{(v-b)^2}+\frac{A_3+B_3 T+C_3 e^{(-kT/T_c)}}{(v-b)^3} \\ +\frac{A_4+B_4 T}{(v-b)^4}+\frac{A_5+B_5 T+C_5 e^{(-kT/T_c)}}{(v-b)^5}+(A_6+B_6 T)e^{av} \tag{24}$$

where the constant coefficients are A_i, B_i, C_i, k, b, and α.

Strobridge (1962) suggested an equation of state that was developed for nitrogen properties and used for most cryogenic fluids. This equation combines the B-W-R equation of state with an equation for high density nitrogen suggested by Benedict (1937). These equations have been used successfully for liquid and vapor phases, extending in the liquid phase to the triple-point temperature and the freezing line, and in the vapor phase from 10 to 1000 K, with pressures to 1 GPa. The equation suggested by Strobridge is accurate within the uncertainty of the measured p-v-T data. This equation, as originally reported by Strobridge, is

$$p = RT\rho + \left[Rn_1 T + n_2 + \frac{n_3}{T} + \frac{n_4}{T^2} + \frac{n_5}{T^4} \right] \rho^2 + (Rn_6 T + n_7) \rho^3 + n_8 T \rho^4$$
$$+ \rho^3 \left[\frac{n_9}{T^2} + \frac{n_{10}}{T^3} + \frac{n_{11}}{T^4} \right] \exp(-n_{16} \rho^2) + \rho^5 \left[\frac{n_{12}}{T^2} + \frac{n_{13}}{T^3} + \frac{n_{14}}{T^4} \right] \exp(-n_{16} \rho^2) + n_{15} \rho^6 \tag{25}$$

The 15 coefficients of this equation's linear terms are determined by a least-square fit to experimental data. Hust and Stewart (1966) and Hust and McCarty (1967) give further information on methods and techniques for determining equations of state.

In the absence of experimental data, Van der Waals' principle of corresponding states can predict fluid properties. This principle relates properties of similar substances by suitable reducing factors; i.e., the p-v-T surfaces of similar fluids in a given region are assumed to be of similar shape. The critical point can be used to define reducing parameters to scale the surface of one fluid to the dimensions of another. Modifications of this principle, as suggested by Kamerlingh Onnes, a Dutch cryogenic researcher, have been used to improve correspondence at low pressures. The principle of corresponding states provides useful approximations, and numerous modifications have been reported. More complex treatments for predicting property values, which recognize similarity of fluid properties, are by generalized equations of state. These equations ordinarily allow for adjustment of the p-v-T surface by introduction of parameters. One example (Hirschfelder et al. 1958) allows for departures from the principle of corresponding states by adding two correlating parameters.

CALCULATING THERMODYNAMIC PROPERTIES

While equations of state provide p-v-T relations, a thermodynamic analysis usually requires values for internal energy, enthalpy, and entropy. These properties have been tabulated for many substances, including refrigerants and can be extracted from such tables by interpolating manually or with a suitable computer program. This approach is appropriate for hand calculations and for relatively simple computer models; however, for many computer simulations, the overhead in memory or input and output required to use tabulated data can make this approach unacceptable. For large thermal system simulations or complex analyses, it may be more efficient to determine internal energy, enthalpy, and entropy using fundamental thermodynamic relations or curves fit to experimental data. Some of these relations are discussed in the following sections. Also, the thermodynamic relations discussed in those sections are the basis for constructing tables of thermodynamic property data. Further information on the topic may be found in references covering system modeling and thermodynamics (Stoecker 1989, Howell and Buckius 1992).

At least two intensive properties must be known to determine the remaining properties. If two known properties are either p, v, or T (these are relatively easy to measure and are commonly used in simulations), the third can be determined throughout the range of interest using an equation of state. Furthermore, if the specific heats at zero pressure are known, specific heat can be accurately determined from spectroscopic measurements using statistical mechanics (NASA 1971). Entropy may be considered a function of T and p, and from calculus an infinitesimal change in entropy can be written as follows:

$$ds = \left(\frac{\partial s}{\partial T}\right)_p dT + \left(\frac{\partial s}{\partial p}\right)_T dp \tag{26}$$

Likewise, a change in enthalpy can be written as

$$dh = \left(\frac{\partial h}{\partial T}\right)_p dT + \left(\frac{\partial h}{\partial p}\right)_T dp \tag{27}$$

Using the relation $Tds = dh - vdp$ and the definition of specific heat at constant pressure, $c_p \equiv (\partial h/\partial T)_p$, Equation (27) can be rearranged to yield

$$ds = \frac{c_p}{T}dT + \left[\left(\frac{\partial h}{\partial p}\right)_T - v\right]\frac{dp}{T} \tag{28}$$

Equations (26) and (28) combine to yield $(\partial s/\partial T)_p = c_p/T$. Then, using the Maxwell relation $(\partial s/\partial p)_T = -(\partial v/\partial T)_p$, Equation (26) may be rewritten as

$$ds = \frac{c_p}{T}dT - \left(\frac{\partial v}{\partial T}\right)_p dp \tag{29}$$

This is an expression for an exact derivative, so it follows that

$$\left(\frac{\partial c_p}{\partial p}\right)_T = -T\left(\frac{\partial^2 v}{\partial T^2}\right)_p \tag{30}$$

Integrating this expression at a fixed temperature yields

$$c_p = c_{p0} - \int_0^p T\left(\frac{\partial^2 v}{\partial T^2}\right) dp_T \tag{31}$$

where c_{p0} is the known zero pressure specific heat, and dp_T is used to indicate that the integration is performed at a fixed temperature. The second partial derivative of specific volume with respect to temperature can be determined from the equation of state. Thus, Equation (31) can be used to determine the specific heat at any pressure.

Using $Tds = dh - vdp$, Equation (29) can be written as

$$dh = c_p dT + \left[v - T\left(\frac{\partial v}{\partial T}\right)_p\right] dp \tag{32}$$

Equations (28) and (32) may be integrated at constant pressure to obtain

part I Theory

$$s(T_1, p_0) = s(T_0, p_0) + \int_{T_0}^{T_1} \frac{c_p}{T} dT_p \tag{33}$$

$$\text{and} \quad h(T_1, p_0) = h(T_0, p_0) + \int_{T_0}^{T_1} c_p dT \tag{34}$$

Integrating the Maxwell relation $(\partial s/\partial p)_T = -(\partial v/\partial T)_p$ gives an equation for entropy changes at a constant temperature as

$$s(T_0, p_1) = s(T_0, p_0) - \int_{p_0}^{p_1} \left(\frac{\partial v}{\partial T}\right)_p dp_T \tag{35}$$

Likewise, integrating Equation (32) along an isotherm yields the following equation for enthalpy changes at a constant temperature

$$h(T_0, p_1) = h(T_0, p_0) + \int_{p_0}^{p_1} \left[v - T\left(\frac{\partial v}{\partial T}\right)_p\right] dp \tag{36}$$

Internal energy can be calculated from $u = h - pv$.

Combinations (or variations) of Equations (33) through (36) can be incorporated directly into computer subroutines to calculate properties with improved accuracy and efficiency. However, these equations are restricted to situations where the equation of state is valid and the properties vary continuously. These restrictions are violated by a change of phase such as evaporation and condensation, which are essential processes in air-conditioning and refrigerating devices. Therefore, the Clapeyron equation is of particular value; for evaporation or condensation it gives

$$\left(\frac{dp}{dT}\right)_{sat} = \frac{s_{fg}}{v_{fg}} = \frac{h_{fg}}{T v_{fg}} \tag{37}$$

where

s_{fg} = entropy of vaporization

h_{fg} = enthalpy of vaporization

v_{fg} = specific volume difference between vapor and liquid phases

If vapor pressure and liquid and vapor density data are known at saturation, and these are relatively easy measurements to obtain, then changes in enthalpy and entropy can be calculated using Equation (37).

Phase Equilibria for Multicomponent Systems

To understand phase equilibria, consider a container full of a liquid made of two components; the more volatile component is designated i and the less volatile component j (Figure 2A). This mixture is all liquid because the temperature is low—but not so low that a solid appears. Heat added at a constant pressure raises the temperature of the mixture, and a sufficient increase causes vapor to form, as shown in Figure 2B. If heat at constant pressure continues to be added, eventually the temperature will become so high that only vapor remains in the container (Figure 2C). A temperature-concentration (T-x) diagram is useful for exploring details of this situation.

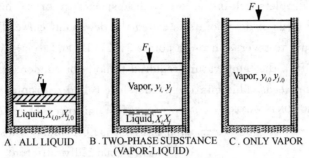

Fig. 2 Mixture of i and j Components in Constant Pressure Container

Figure 3 is a typical T-x diagram valid at a fixed pressure. The case shown in Figure 2A, a container full of liquid mixture with mole fraction $x_{i,0}$ at temperature T_0, is point 0 on the T-x diagram. When heat is added, the temperature of the mixture increases. The point at which vapor begins to form is the **bubble point**. Starting at point 0, the first bubble will form at temperature T_1, designated by point 1 on the diagram. The locus of bubble points is the **bubble point curve**, which provides bubble points for various liquid mole fractions x_i.

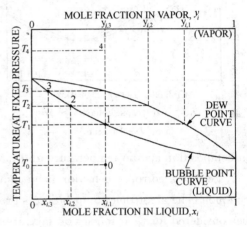

Fig. 3 Temperature-Concentration(T-x) Diagram for Zeotropic Mixture

When the first bubble begins to form, the vapor in the bubble may not have the i mole fraction found in the liquid mixture. Rather, the mole fraction of the more volatile species is higher in the vapor than in the liquid. Boiling prefers the more volatile species, and the T-x diagram shows this behavior. At T_1, the vaporforming bubbles have an i mole fraction of $y_{i,1}$. If heat continues to be added, this preferential boiling will deplete the liquid of species i and the temperature required to continue the process will increase. Again, the T-x diagram reflects this fact; at point 2 the i mole fraction in the liquid is reduced to $x_{i,2}$ and the vapor has a mole fraction of $y_{i,2}$. The temperature required to boil the mixture is increased to T_2. Position 2 on the T-x diagram could correspond to the physical situation shown in Figure 2B.

If the constant-pressure heating continues, all the liquid eventually becomes vapor at temperature T_3. At this point the i mole fraction in the vapor $y_{i,3}$ equals the starting mole fraction in the all-liquid mixture $x_{i,1}$. This equality is required for mass and species conservation. Further addition of heat simply raises the vapor temperature. The final position 4 corresponds to the physical situation shown in Figure 2C.

Starting at position 4 in Figure 3, the removal of heat leads to 3, and further heat removal would cause droplets rich in the less volatile species to form. This point is called the **dew point**, and the locus of dew points is called the **dew-point curve**. The removal of heat will cause the mixture to reverse through points 3, 2, 1, and to starting point 0. Because the composition shifts, the temperature required to boil (or condense) this mixture changes as the process proceeds. This mixture is therefore called **zeotropic**.

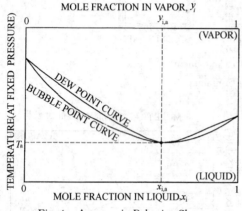

Fig. 4 Azeotropic Behavior Shown on T-x Diagram

Most mixtures have T-x diagrams that behave as previously described, but some have a markedly different feature. If the dew point and bubble point curves intersect at any point other than at their ends, the mixture exhibits what is called **azeotropic** behavior at that composition. This case is shown as position a in the T-x diagram of Figure 4. If a container of liquid with a mole fraction x_a were boiled, vapor would be formed with an identical mole fraction y_a. The addition of heat at constant pressure would continue with no shift in composition and no temperature glide.

Perfect azeotropic behavior is uncommon, while near azeotropic behavior is fairly common. The azeotropic composition is pressure dependent, so operating pressures should be considered for their impact on mixture behavior. Azeotropic and near-azeotropic refrigerant mixtures find wide application. The properties of an azeotropic mixture are such that they may be conveniently treated as pure substance properties. Zeotropic mixtures, however, require special treatment, using an equation-of-state approach with appropriate mixing rules or using the fugacities with the standard state method (Tassios 1993). Refrigerant and lubricant blends are a zeotropic mixture and can be treated by these methods (see Thome 1995 and Martz et al. 1996a, b).

COMPRESSION REFRIGERATION CYCLES

CARNOT CYCLE

The Carnot cycle, which is completely reversible, is a perfect model for a refrigeration cycle operating between two fixed temperatures, or between two fluids at different temperatures and each with infinite heat capacity. Reversible cycles have two important properties: (1) no refrigerating cycle may have a coefficient of performance higher than that for a reversible cycle operated between the same temperature limits, and (2) all reversible cycles, when operated between the same temperature limits, have the same coefficient of performance. Proof of both statements may be found in almost any textbook on elementary engineering thermodynamics.

Figure 5 shows the Carnot cycle on temperature-entropy coordinates. Heat is withdrawn at the constant temperature T_R from the region to be refrigerated. Heat is rejected at the constant ambient temperature T_0. The cycle is completed by an isentropic expansion and an isentropic compression. The energy transfers are given by

$$Q_0 = T_0(S_2 - S_3)$$
$$Q_i = T_R(S_1 - S_4) = T_R(S_2 - S_3)$$
$$W_{net} = Q_0 - Q_i$$

Thus, by Equation (15),

$$\text{COP} = \frac{T_R}{T_0 - T_R} \tag{38}$$

Example 1. Determine entropy change, work, and coefficient of performance for the cycle shown in Figure 6. Temperature of the refrigerated space T_R is 250 K and that of the atmosphere T_0 is 300 K. Refrigeration load is 125 kJ.

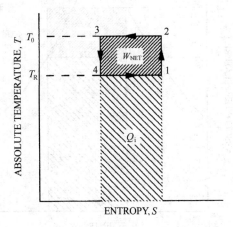

Fig. 5 Carnot Refrigeration Cycle

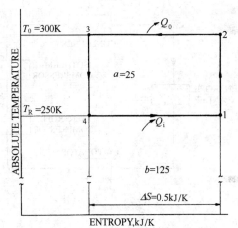

Fig. 6 Temperature-Entropy Diagram for Carnot Refrigeration Cycle of Example 1

Solution:

$$\Delta S = S_1 - S_4 = Q_i / T_R = 125/250 = 0.5 \text{ kJ/K}$$
$$W = \Delta S(T_0 - T_R) = 0.5(300 - 250) = 25 \text{ kJ}$$
$$\text{COP} = Q_i / (Q_0 - Q_i) = Q_i / W = 125/25 = 5$$

Flow of energy and its area representation in Figure 6 is:

Energy	kJ	Area
Q_i	125	b
Q_0	150	a+b
W	25	a

part I Theory

The net change of entropy of any refrigerant in any cycle is always zero. In Example 1 the change in entropy of the refrigerated space is $\Delta S_R = -125/250 = -0.5 \text{kJ/K}$ and that of the atmosphere is $\Delta S_0 = 125/250 = 0.5 \text{kJ/K}$. The net change in entropy of the isolated system is $\Delta S_{total} = \Delta S_R + \Delta S_0 = 0$.

The Carnot cycle in Figure 7 shows a process in which heat is added and rejected at constant pressure in a two-phase region of a refrigerant. Saturated liquid at state 3 expands isentropically to the low temperature and pressure of the cycle at state d. Heat is added isothermally and isobarically by evaporating the liquid phase refrigerant from state d to state 1. The cold saturated vapor at state 1 is compressed isentropically to the high temperature in the cycle at state b. However the pressure at state b is below the saturation pressure corresponding to the high temperature in the cycle. The compression process is completed by an isothermal compression process from state b to state c. The cycle is completed by an isothermal and isobaric heat rejection or condensing process from state c to state 3.

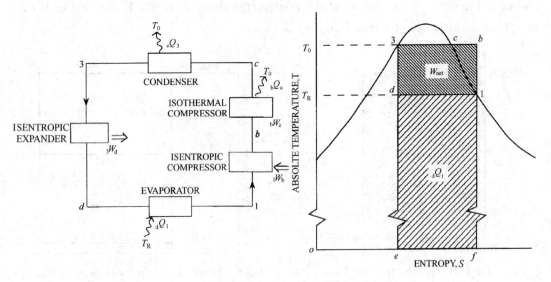

Fig. 7 Carnot Vapor Compression Cycle

Applying the energy equation for a mass of refrigerant m yields (all work and heat transfer are positive)

$$_3W_d = m(h_3 - h_d)$$

$$_1W_b = m(h_b - h_1)$$

$$_bW_c = T_0(S_b - S_c) - m(h_b - h_c)$$

$$_dQ_1 = m(h_1 - h_d) = \text{Area defld}$$

The net work for the cycle is

$$W_{net} = {}_1W_b + {}_bW_c - {}_3W_d = \text{Area d1bc3d}$$

and

$$\text{COP} = \frac{{}_dQ_1}{W_{net}} = \frac{T_R}{T_0 - T_R}$$

THEORETICAL SINGLE-STAGE CYCLE USING A PURE REFRIGERANT OR AZEOTROPIC MIXTURE

A system designed to approach the ideal model shown in Figure 7 is desirable. A pure refrigerant or an azeotropic mixture can be used to maintain constant temperature during the phase changes by maintaining a constant pressure. Because of such concerns as high initial cost and increased maintenance requirements, a practical machine has one compressor instead of two and the expander (engine or turbine) is replaced by a simple expansion valve. The valve throttles the refrigerant from high pressure to low pressure. Figure 8 shows the theoretical single-stage cycle used as a model for actual systems.

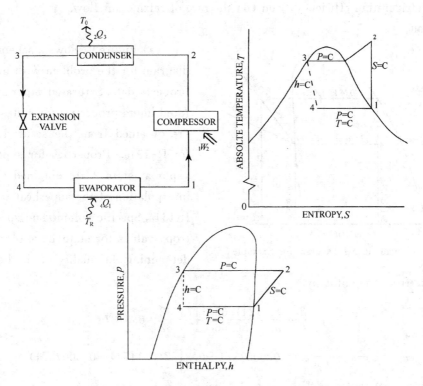

Fig. 8 Theoretical Single-Stage Vapor Compression Refrigeration Cycle

Applying the energy equation for a mass of refrigerant m yields

$$_4Q_1 = m(h_1 - h_4)$$

$$_1W_2 = m(h_2 - h_1)$$

$$_2Q_3 = m(h_2 - h_3)$$

$$h_3 = h_4 \tag{39}$$

The constant enthalpy throttling process assumes no heat transfer or change in potential or kinetic energy through the expansion valve.

The coefficient of performance is

$$\text{COP} = \frac{_4Q_1}{_1W_2} = \frac{h_1 - h_4}{h_2 - h_1} \tag{40}$$

The theoretical compressor displacement CD (at 100% volumetric efficiency), is

$$\text{CD} = \dot{m}v_3 \tag{41}$$

which is a measure of the physical size or speed of the compressor required to handle the prescribed refrigeration load.

Example 2. A theoretical single-stage cycle using R-134a as the refrigerant operates with a condensing temperature of 30℃ and an evaporating temperature of −20℃. The system produces 50 kW of refrigeration. Determine (a) the thermodynamic property values at the four main state points of the cycle, (b) the coefficient of performance of the cycle, (c) the cycle refrigerating efficiency, and (d) the rate of refrigerant flow.

Solution:

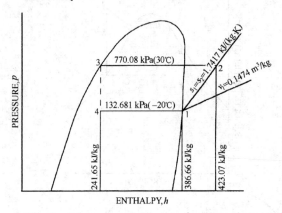

Fig. 9 Schematic p-h Diagram for Example 2

(a) Figure 9 shows a schematic p-h diagram for the problem with numerical property data. Saturated vapor and saturated liquid properties for states 1 and 3 are obtained from the saturation table for R-134a. Properties for superheated vapor at state 2 are obtained by linear interpolation of the superheat tables for R-134a. Specific volume and specific entropy values for state 4 are obtained by detemining the quality of the liquid-vapor mixture from the enthalpy.

$$x_4 = \frac{h_4 - h_f}{h_g - h_f} = \frac{241.65 - 173.82}{386.66 - 173.82} = 0.3187$$

$$v_4 = v_f + x_4(v_g - v_f) = 0.0007374 + 0.3187(0.14744 - 0.0007374)$$

$$= 0.04749 \text{m}^3/\text{kg}$$

$$s_4 = s_f + x_4(s_g - s_f) = 0.9009 + 0.3187(1.7417 - 0.9009)$$

$$= 1.16886 \text{kJ}/(\text{kg} \cdot \text{K})$$

The property data are tabulated in Table 1.

Thermodynamic Property Data for Example 2   Table 1

State	t, ℃	p, kPa	v, m³/kg	h, kJ/kg	s, kJ/(kg·K)
1	−20.0	132.68	0.14744	386.66	1.7417
2	37.8	770.08	0.02798	423.07	1.7417
3	30.0	770.08	0.00084	241.65	1.1432
4	−20.0	132.68	0.04749	241.65	1.1689

(b) By Equation (40)

$$\text{COP} = \frac{386.66 - 241.65}{423.07 - 386.66} = 3.98$$

(c) By Equation (17)

$$\eta_R = \frac{\text{COP}(T_3 - T_1)}{T_1} = \frac{(3.98)(50)}{253.15} = 0.79 \text{ or } 79\%$$

(d) The mass flow of refrigerant is obtained from an energy balance on the evaporator. Thus

$$\dot{m}(h_1 - h_4) = q_i = 50 \text{kW}$$

$$\text{and } \dot{m} = \frac{\dot{Q}_i}{(h_1 - h_4)} = \frac{50}{(386.66 - 241.65)} = 0.345 \text{kg/s}$$

The saturation temperatures of the single-stage cycle have a strong influence on the magnitude of the coefficient of performance. This influence may be readily appreciated by an area analysis on a temperature-entropy (T-S) diagram. The area under a reversible process line on a T-S diagram is directly proportional to the thermal energy added or removed from the working fluid. This observation follows directly from the definition of entropy [see Equation (8)].

In Figure 10 the area representing Q_0 is the total area under the constant pressure curve between states 2 and 3. The area representing the refrigerating capacity Q_i is the area under the constant pressure line connecting states 4 and 1. The net work required W_{net} equals the difference ($Q_0 - Q_i$), which is represented by the shaded area shown on Figure 10.

Because COP = Q_i/W_{net}, the effect on the COP of changes in evaporating temperature and condensing temperature may be observed. For example, a decrease in evaporating temperature T_E significantly increases W_{net} and slightly decreases Q_i. An increase in condensing temperature T_C produces the same results but with less effect on W_{net}.

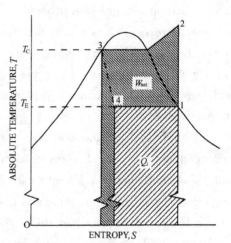

Fig. 10   Areas on T-S Diagram Representing Refrigerating Effect and Work Supplied for Theoretical Single-Stage Cycle

Therefore, for maximum coefficient of performance, the cycle should operate at the lowest possible condensing temperature and at the maximum possible evaporating temperature.

LORENZ REFRIGERATION CYCLE

The Carnot refrigeration cycle includes two assumptions which make it impractical. The heat transfer capacity of the two external fluids are assumed to be infinitely large so the external fluid temperatures remain fixed at T_0 and T_R (they become infinitely large thermal reservoirs). The Carnot cycle also has no thermal resistance between the working refrigerant and the external fluids in the two heat exchange processes. As a result, the refrigerant must remain fixed at T_0 in the condenser and at T_R in the evaporator.

The Lorenz cycle eliminates the first restriction in the Carnot cycle and allows the temperature of the two external fluids to vary during the heat exchange. The second assumption of negligible thermal resistance between the working refrigerant and the two external fluids remains. Therefore the refrigerant temperature must change during the two heat exchange processes to equal the changing temperature of the external fluids. This cycle is completely reversible when operating between two fluids, each of which has a finite but constant heat capacity.

Figure 11 is a schematic of a Lorenz cycle. Note that this cycle does not operate between two fixed temperature limits. Heat is added to the refrigerant from state 4 to state 1. This process is assumed to be linear on T-S coordinates, which represents a fluid with constant heat capacity. The temperature of the refrigerant is increased in an isentropic compression process from state 1 to state 2. Process 2-3 is a heat rejection process in which the refrigerant temperature decreases linearly with heat transfer. The cycle is concluded with an isentropic expansion process between states 3 and 4.

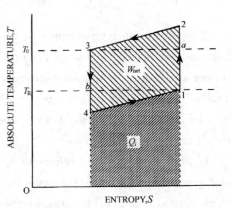

Fig. 11 Processes of Lorenz Refrigeration Cycle

The heat addition and heat rejection processes are parallel so the entire cycle is drawn as a parallelogram on T-S coordinates. A Carnot refrigeration cycle operating between T_0 and T_R would lie between states 1, a, 3, and b. The Lorenz cycle has a smaller refrigerating effect than the Carnot cycle and more work is required. However this cycle is a more practical reference to use than the Carnot cycle when a refrigeration system operates between two single phase fluids such as air or water.

The energy transfers in a Lorenz refrigeration cycle are as follows, where ΔT is the temperature change of the refrigerant during each of the two heat exchange processes.

$$Q_0 = (T_0 + \Delta T/2)(S_2 - S_3)$$

$$Q_i = (T_R - \Delta T/2)(S_1 - S_4) = (T_R - \Delta T/2)(S_2 - S_3)$$

$$W_{net} = Q_0 - Q_R$$

Thus by Equation (15),

$$\text{COP} = \frac{T_R - (\Delta T/2)}{T_0 - T_R + \Delta T} \tag{42}$$

Example 3. Determine the entropy change, the work required, and the coefficient of performance for the Lorenz cycle shown in Figure 11 when the temperature of the refrigerated space is $T_R = 250K$, the ambient temperature is $T_0 = 300K$, the ΔT of the refrigerant is 5K and the refrigeration load is 125 kJ.

Solution:

$$\Delta S = \int_4^1 \frac{Q_i}{T} = \frac{Q_i}{T_R - (\Delta T/2)} = \frac{125}{247.5} = 0.5051 \text{kJ/K}$$

$$Q_0 = [T_0 + (\Delta T/2)] \Delta S = (300 + 2.5)0.5051 = 152.78 \text{kJ}$$

$$W_{net} = Q_0 - Q_R = 152.78 - 125 = 27.78 \text{kJ}$$

$$\text{COP} = \frac{T_R - (\Delta T/2)}{T_0 - T_R + \Delta T} = \frac{250 - (5/2)}{300 - 250 + 5} = \frac{247.5}{55} = 4.50$$

Note that the entropy change for the Lorenz cycle is larger than for the Carnot cycle at the same temperature levels and the same capacity (see Example 1). That is, the heat rejection is larger and the work requirement is also larger for the Lorenz cycle. This difference is caused by the finite temperature difference between the working fluid in the cycle compared to the bounding temperature reservoirs. However, as discussed previously, the assumption of constant temperature heat reservoirs is not necessarily a good representation of an actual refrigeration system because of the temperature changes that occur in the heat exchangers.

THEORETICAL SINGLE-STAGE CYCLE USING ZEOTROPIC REFRIGERANT MIXTURE

A practical method to approximate the Lorenz refrigeration cycle is to use a fluid mixture as the refrigerant and the four system components shown in Figure 8. When the mixture is not azeotropic and the phase change processes occur at constant pressure, the temperatures change during the evaporation and condensation processes and the theoretical single-stage cycle can be shown on T-S coordinates as in Figure 12. This can be compared with Figure 10 in which the system is shown operating with a pure simple substance or an azeotropic mixture as the refrigerant. Equations (14), (15), (39), (40), and (41) apply to this cycle and to conventional cycles with constant phase change temperatures. Equation (42) should be used as the reversible cycle COP in Equation (17).

For zeotropic mixtures, the concept of constant saturation temperatures does not exist. For example, in the evaporator, the refrigerant en-

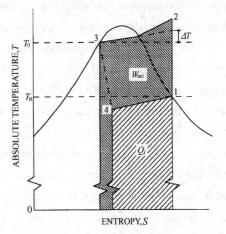

Fig. 12 Areas on T-S Diagram Representing Refrigerating Effect and Work Supplied for Theoretical Single-Stage Cycle Using Zeotropic Mixture as Refrigerant

ters at T_4 and exits at a higher temperature T_1. The temperature of saturated liquid at a given pressure is the **bubble point** and the temperature of saturated vapor at a given pressure is called the **dew point.** The temperature T_3 on Figure 12 is at the bubble point at the condensing pressure and T_1 is at the dew point at the evaporating pressure.

An analysis of areas on a $T\text{-}s$ diagram representing additional work and reduced refrigerating effect from a Lorenz cycle operating between the same two temperatures T_1 and T_3 with the same value for ΔT can be performed. The cycle matches the Lorenz cycle most closely when counterflow heat exchangers are used for both the condenser and the evaporator.

In a cycle that has heat exchangers with finite thermal resistances and finite external fluid capacity rates, Kuehn and Gronseth (1986) showed that a cycle which uses a refrigerant mixture has a higher coefficient of performance than a cycle that uses a simple pure substance as a refrigerant. However, the improvement in COP is usually small. The performance of the cycle that uses a mixture can be improved further by reducing the thermal resistance of the heat exchangers and passing the fluids through them in a counterflow arrangement.

MULTISTAGE VAPOR COMPRESSION REFRIGERATION CYCLES

Multistage vapor compression refrigeration is used when several evaporators are needed at various temperatures such as in a supermarket or when the temperature of the evaporator becomes very low. Low evaporator temperature indicates low evaporator pressure and low refrigerant density into the compressor. Two small compressors in series have a smaller displacement and usually operate more efficiently than one large compressor that covers the entire pressure range from the evaporator to the condenser. This is especially true in refrigeration systems that use ammonia because of the large amount of superheating that occurs during the compression process.

The thermodynamic analysis of multistage cycles is similar to the analysis of single-stage cycles. The main difference is that the mass flow differs through various components of the system. A careful mass balance and energy balance performed on individual components or groups of components ensures the correct application of the first law of thermodynamics. Care must also be exercised when performing second law calculations. Often the refrigerating load is comprised of more than one evaporator, so the total system capacity is the sum of the loads from all evaporators. Likewise the total energy input is the sum of the work into all compressors. For multistage cycles, the expression for the coefficient of performance given in Equation 15 should be written as

$$\text{COP} = \Sigma Q_i / W_{net} \tag{43}$$

When compressors are connected in series, the vapor between stages should be cooled to bring the vapor to saturated conditions before proceeding to the next stage of compression. Intercooling usually minimizes the displacement of the compressors, reduces the work requirement, and increases the COP of the cycle. If the refrigerant temperature between stages is above ambient, a simple intercooler that removes heat from the refrigerant can be used. If the temperature is below ambient, which is the usual case, the refrigerant itself must be used to cool the vapor. This is accomplished with a flash intercooler. Figure 13 shows a cycle with a flash intercooler installed.

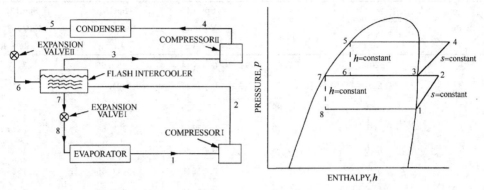

Fig. 13　Schematic and Pressure-Enthalpy Diagram for Dual-Compression, Dual-Expansion Cycle of Example 4

The superheated vapor from compressor Ⅰ is bubbled through saturated liquid refrigerant at the intermediate pressure of the cycle. Some of this liquid is evaporated when heat is added from the superheated refrigerant. The result is that only saturated vapor at the intermediate pressure is fed to compressor Ⅱ. A common assumption is to operate the intercooler at about the geometric mean of the evaporating and condensing pressures. This operating point provides the same pressure ratio and nearly equal volumetric efficiencies for the two compressors. Example 4 illustrates the thermodynamic analysis of this cycle.

Example 4. Determine the thermodynamic properties of the eight state points shown in Figure 13, the mass flows, and the COP of this theoretical multistage refrigeration cycle when R-134a is the refrigerant. The saturated evaporator temperature is $-20℃$, the saturated condensing temperature is $30℃$, and the refrigeration load is 50kW. The saturation temperature of the refrigerant in the intercooler is $0℃$, which is nearly at the geometric mean pressure of the cycle.

Solution:

Thermodynamic property data are obtained from the saturation and superheat tables for R-134a. States 1, 3, 5, and 7 are obtained directly from the saturation table. State 6 is a mixture of liquid and vapor. The quality is calculatied by

$$x_6 = \frac{h_6 - h_7}{h_3 - h_7} = \frac{241.65 - 200}{398.68 - 200} = 0.20963$$

Then,

$$v_6 = v_7 + x_6(v_3 - v_7) = 0.000773 + 0.20963(0.06935 - 0.000773) = 0.01515 \text{m}^3/\text{kg}$$

$$s_6 = s_7 + x_6(s_3 - s_7) = 1.0 + 0.20963(1.7274 - 1.0) = 1.15248 \text{kJ/(kg·K)}$$

Similarly for state 8.

$$x_8 = 0.12300, \quad v_8 = 0.01878 \text{m}^3/\text{kg}, \quad s_8 = 1.0043 \text{kJ/(kg·K)}$$

States 2 and 4 are obtained from the superheat tables by linear interpolation. The thermodynamic property data are summarized in Table 2.

Thermodynamic Property Values for Example 4   Table 2

State	Temperature, ℃	Pressure, kPa	Specific Volume, m³/kg	Specific Enthalpy, kJ/kg	Specific Entropy, kJ/(kg·K)
1	−20.0	132.68	0.14744	386.66	1.7417
2	2.8	292.69	0.07097	401.51	1.7417
3	0.0	292.69	0.06935	398.68	1.7274
4	33.6	770.08	0.02726	418.68	1.7274
5	30.0	770.08	0.00084	241.65	1.1432
6	0.0	292.69	0.01515	241.65	1.1525
7	0.0	292.69	0.00077	200.00	1.0000
8	−20.0	132.68	0.01878	200.00	1.0043

The mass flow through the lower circuit of the cycle is determined from an energy balance on the evaporator.

$$\dot{m}_1 = \frac{\dot{Q}_i}{h_1 - h_8} = \frac{50}{386.66 - 200} = 0.2679 \text{kg/s}$$

$$\dot{m}_1 = \dot{m}_2 = \dot{m}_7 = \dot{m}_8$$

For the upper circuit of the cycle,

$$\dot{m}_3 = \dot{m}_4 = \dot{m}_5 = \dot{m}_6$$

Assuming the intercooler has perfect external insulation, an energy balance on it is used to compute $\dot{m}_3$.

$$\dot{m}_6 h_6 + \dot{m}_2 h_2 = \dot{m}_7 h_7 + \dot{m}_3 h_3$$

Rearranging and solving for $\dot{m}_3$,

$$\dot{m}_3 = \dot{m}_2 \frac{h_7 - h_2}{h_6 - h_3} = 0.2679 \frac{200 - 401.51}{241.65 - 398.68} = 0.3438 \text{kg/s}$$

$$\dot{W}_I = \dot{m}_1(h_2 - h_1) = 0.2679(401.51 - 386.66) = 3.978 \text{kW}$$

$$\dot{W}_{II} = \dot{m}_3(h_4 - h_3) = 0.3438(418.68 - 398.68) = 6.876 \text{kW}$$

$$\text{COP} = \frac{\dot{Q}_i}{\dot{W}_I + \dot{W}_{II}} = \frac{50}{3.978 + 6.876} = 4.61$$

Examples 2 and 4 have the same refrigeration load and operate with the same evaporating and condensing temperatures. The two-stage cycle in Example 4 has a higher COP and less work input than the single-stage cycle. Also the highest refrigerant temperature leaving the compressor is about 34℃ for the two-stage cycle versus about 38℃ for the single-stage cycle. These differences are more pronounced for cycles operating at larger pressure ratios.

WORDS AND EXPRESSIONS

account for	v. 说明，占，解决，得分
ammonia	n. ［化］氨，氨水
azeotropic	adj. ［化］共沸的，恒沸点的
bound	n. 跃进，跳，范围，限度；adj. 正要启程的，开往…去的，被束缚的，装订的；v. 跳跃，限制
bubble point	起泡点，始沸点
Carnot cycle	卡诺循环(Lorenz cycle 劳伦兹循环)
cascade	n. 小瀑布，喷流；vi. 成瀑布落下；n. 层叠
cocurrent	n. 直流；同向
cohesive	adj. 黏着的，有黏聚性的
counterflow	n. 逆流
entropy	n. ［物］熵，［无］平均信息量
equation of state	状态方程
equilibria	n. 均衡，均势(eguilibrium 的复数形式)
expansion valve	膨胀阀
fluorinate	v. 使与氟素化合，在(饮水)加少量之氟
heat rejection	排热，热损失(heat addition 供热)
in the absence of	缺乏…时，当…不在时
initial cost	初投资
internal combustion engine	内燃机
interpolation	n. 篡改，添写，插补
multistage	adj. 多级的
neutron	n. 中子
nuclear (atomic) energy	核能(原子能)
proton	n. ［核］质子
reference plane	基准面，参考面
reversible	adj. 可逆的
serve as	适于
shaded area	阴影面积
shaft	n. 轴，杆状物
shift	n. 移动，轮班，移位，变化，办法，手段；vt. 替换，转移，改变，移转，推卸，变速；vi. 转换，移动，转变，推托，变速
single-stage cycle	单级循环(multistage cycle 多级循环)
specific volume	比容
stationary	adj. 固定的

tabulate	*vt.* 把...制成表格；*v.* 列表
the product of	......的乘积
throttle	*v.* 节流，扼杀
virial	*n.* ［核］维里（作用于粒子上的合力与粒子矢径的标积）
volatile	*adj.* 飞行的，挥发性的，可变的，不稳定的，轻快的，爆炸性的
withdraw	*vt.* 收回，撤销；*vi.* 缩回，退出；*v.* 撤退

NOTATIONS

(1) In general, a truncated virial expansion of four terms is valid for densities of less than one-half the value at the critical point.

一般说来，截断的维里展开四项式对小于临界点密度一半值的密度是正确的。

(2) The 15 coefficients of this equation's linear terms are determined by a least-square fit to experimental data.

这个方程线性项的 15 个系数是由实验数据的最小二乘方拟合确定。

EXERCISES

请翻译下列句子。

1) Work is the mechanism that transfers energy across the boundary of systems with differing pressures (of force of any kind), always toward the lower pressure. If the total effect produced in the system can be reduced to the raising of a weight, then nothing but work has crossed the boundary. Work is positive then energy is removed from the system.

2) A process is described by specifying the initial and final equilibrium states, the path (if identifiable), and the interactions that take place across system boundaries during the process.

3) The concept of irreversibility provides added insight into the operation of cycles. For example, the larger the irreversibility in a refrigeration cycle operating with a given refrigeration load between two fixed temperature levels, the larger the amount of work required to operate the cycle. Irreversibilities include pressure drops in lines and heat exchangers, heat transfer between fluids of different temperature, and mechanical friction. Reducing total irreversibility in a cycle improves the cycle performance. In the limit of no irreversibilities, a cycle will attain its maximum ideal efficiency.

4) Equation (11) is commonly applied to a system with one mass flow in, the same mass flow out, no work, and negligible kinetic or potential energy flows.

5) In a cycle, the reduction of work produced by a power cycle or the increase in work required by a refrigeration cycle is equal to the absolute ambient temperature multiplied by the sum of the irreversibilities in all the processes in the cycle.

6) COP is defined as the benefit of the cycle (amount of heat removed) divided by the required energy input to operate the cycle, or

$$\text{COP} \equiv \frac{\text{Useful refrigerationg effect}}{\text{Net energy supplied from external sources}}$$

7) This approach is appropriate for hand calculations and for relatively simple computer models; however, for many computer simulations, the overhead in memory or input and output required to use tabulated data can make this approach unacceptable. For large thermal system simulations or complex analyses, it may be more efficient to determine internal energy, enthalpy, and entropy using fundamental thermodynamic relations or curves fit to experimental data.

8) Figure 5 shows the Carnot cycle on temperature-entropy coordinates. Heat is withdrawn at the constant temperature T_R from the region to be refrigerated. Heat is rejected at the constant ambient temperature T_0. The cycle is completed by an isentropic expansion and an isentropic compression.

LESSON 2

FLUID FLOW

FLOWING fluids in heating, ventilating, air-conditioning, and refrigeration systems transfer heat and mass. This chapter introduces the basics of fluid mechanics that are related to HVAC processes, reviews pertinent flow processes, and presents a general discussion of single-phase fluid flow analysis.

FLUID PROPERTIES

Fluids differ from solids in their reaction to shearing. When placed under shear stress, a solid deforms only a finite amount, whereas a fluid deforms continuously for as long as the shear is applied. Both liquids and gases are fluids. Although liquids and gases differ strongly in the nature of molecular actions, their primary mechanical differences are in the degree of compressibility and liquid formation of a free surface (interface).

Fluid motion can usually be described by one of several simplified modes of action or models. The simplest is the ideal-fluid model, which assumes no resistance to shearing. Ideal flow analysis is well developed (Baker 1983, Schlichting 1979, Streeter and Wylie 1979), and when properly interpreted is valid for a wide range of applications. Nevertheless, the effects of viscous action may need to be considered. Most fluids in HVAC applications can be treated as Newtonian, where the rate of deformation is directly proportional to the shearing stress. Turbulence complicates fluid behavior, and viscosity influences the nature of the turbulent flow.

Density

The density ρ of a fluid is its mass per unit volume. The densities of air and water at standard indoor conditions of 20°C and 101.325 kPa (sea level atmospheric pressure) are

$$\rho_{water} = 998 \text{kg/m}^3$$
$$\rho_{air} = 1.20 \text{kg/m}^3$$

Viscosity

Viscosity is the resistance of adjacent fluid layers to shear. For shearing between two parallel plates, each of area A and separated by distance Y, the tangential force F per unit

---

Extracted fron Chapter 2 of the ASHRAE *Handbook-Fundamentals*.

area required to slide one plate with velocity V parallel to the other is proportional to V/Y:
$$F/A = \mu(V/Y)$$
where the proportionality factor μ is the **absolute viscosity** or **dynamic viscosity** of the fluid. The ratio of the tangential force F to area A is the **shearing stress** τ, and V/Y is the **lateral velocity gradient** (Figure 1A). In complex flows, velocity and shear stress may vary across the flow field; this is expressed by the following differential equation:

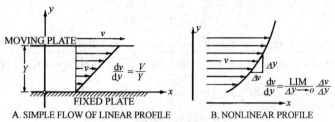

A. SIMPLE FLOW OF LINEAR PROFILE   B. NONLINEAR PROFILE
Fig. 1   Velocity Profiles and Gradients in Shear Flows

$$\tau = \mu \frac{dv}{dy} \tag{1}$$

The velocity gradient associated with viscous shear for a simple case involving flow velocity in the x direction but of varying magnitude in the y direction is illustrated in Figure 1B.

Absolute viscosity μ depends primarily on temperature. For gases (except near the critical point), viscosity increases with the square root of the absolute temperature, as predicted by the kinetic theory. Liquid viscosity decreases with increasing temperature.

Absolute viscosity has dimensions of force · time/length^2. At standard indoor conditions, the absolute viscosities of water and dry air are

$$\mu_{\text{water}} = 1.0 \, \text{mN} \cdot \text{s/m}^2$$
$$\mu_{\text{air}} = 18 \, \mu\text{N} \cdot \text{s/m}^2$$

In fluid dynamics, **kinematic viscosity** ν is the ratio of absolute viscosity to density:
$$\nu = \mu/\rho$$
At standard indoor conditions, the kinematic viscosities of water and dry air are
$$\nu_{\text{water}} = 1.00 \times 10^{-6} \, \text{m}^2/\text{s}$$
$$\nu_{\text{air}} = 16 \times 10^{-6} \, \text{m}^2/\text{s}$$

BASIC RELATIONS OF FLUID DYNAMICS

This section considers homogeneous, constant-property, incompressible fluids and introduces fluid dynamic considerations used in most analyses.

Continuity

Conservation of matter applied to fluid flow in a conduit requires that

$$\int \rho v \, dA = \text{constant}$$

where

v = velocity normal to the differential area dA

ρ = fluid density

Both ρ and v may vary over the cross section A of the conduit. If both ρ and v are constant over the cross-sectional area normal to the flow, then

$$\dot{m} = \rho V A = \text{constant} \tag{2a}$$

where $\dot{m}$ is the mass flow rate across the area normal to the flow. When flow is effectively incompressible, ρ = constant; in pipeline and duct flow analyses, the average velocity is then $V = (1/A)\int v\,dA$. The continuity relation is

$$Q = AV = \text{constant} \tag{2b}$$

where Q is the volumetric flow rate. Except when branches occur, Q is the same at all sections along the conduit.

For the ideal-fluid model, flow patterns around bodies (or in conduit section changes) result from displacement effects. An obstruction in a fluid stream, such as a strut in a flow or a bump on the conduit wall, pushes the flow smoothly out of the way, so that behind the obstruction, the flow becomes uniform again. The effect of fluid inertia (density) appears only in pressure changes.

Pressure Variation Across Flow

Pressure variation in fluid flow is important and can be easily measured. Variation across streamlines involves fluid rotation (vorticity). Lateral pressure variation across streamlines is given by the following relation (Bober and Kenyon 1980, Olson 1980, Robertson 1965):

$$\frac{\partial}{\partial r}\left(\frac{p}{\rho} + gz\right) = \frac{v^2}{r} \tag{3}$$

where

r = radius of curvature of the streamline

z = elevation

This relation explains the pressure difference found between the inside and outside walls of a bend and near other regions of conduit section change. It also states that pressure variation is hydrostatic ($p + \rho g z$ = constant) across any conduit where streamlines are parallel.

Bernoulli Equation and Pressure Variation along Flow

A basic tool of fluid flow analysis is the Bernoulli relation, which involves the principle of energy conservation along a streamline. Generally, the Bernoulli equation is not applicable across streamlines. The first law of thermodynamics can be applied to mechanical

LESSON 2

flow energies (kinetic and potential) and thermal energies; heat is a form of energy and energy is conserved.

The change in energy content ΔE per unit mass of flowing material is a result from the work W done on the system plus the heat Q absorbed:

$$\Delta E = W + Q$$

Fluid energy is composed of kinetic, potential (due to elevation z), and internal (u) energies. Per unit mass of fluid, the above energy change relation between two sections of the system is

$$\Delta\left(\frac{v^2}{2} + gz + u\right) = E_M - \Delta\left(\frac{p}{\rho}\right) + Q$$

where the work terms are (1) the external work E_M from a fluid machine (E_M is positive for a pump or blower) and (2) the pressure or flow work p/ρ. Rearranging, the energy equation can be written as the **generalized Bernoulli equation**:

$$\Delta\left(\frac{v^2}{2} + gz + \frac{p}{\rho}\right) + \Delta u = E_M + Q \tag{4}$$

The term in parentheses in Equation (4) is the **Bernoulli constant**:

$$\frac{p}{\rho} + \frac{v^2}{2} + gz = B \tag{5a}$$

In cases with no viscous action and no work interaction, B is constant; more generally its change (or lack thereof) is considered in applying the Bernoulli equation. The terms making up B are fluid energies (pressure, kinetic, and potential) per mass rate of fluid flow. Alternative forms of this relation are obtained through multiplication by ρ or division by g:

$$p + \frac{\rho v^2}{2} + \rho g z = \rho B \tag{5b}$$

$$\frac{p}{\rho g} + \frac{v^2}{2g} + z = \frac{B}{g} \tag{5c}$$

The first form involves energies per volume flow rate, or pressures; the second involves energies per mass flow rate, or heads. In gas flow analysis, Equation (5b) is often used with the $\rho g z$ term dropped as negligible. Equation (5a) should be used when density variations occur. For liquid flows, Equation (5c) is commonly used. Identical results are obtained with the three forms if the units are consistent and the fluids are homogeneous.

Many systems of pipes or ducts and pumps or blowers can be considered as one-dimensional flow. The Bernoulli equation is then considered as velocity and pressure vary along the conduit. Analysis is adequate in terms of the section-average velocity V of Equation (2a) or (2b). In the Bernoulli relation [Equations (4) and (5)], v is replaced by V, and variation across streamlines can be ignored; the whole conduit is now taken as one streamline. Two- and three-dimensional details of local flow occurrences are still significant, but their effect is combined and accounted for in factors.

The kinetic energy term of the Bernoulli constant B is expressed as $\alpha V^2/2$, where the

part I *Theory*

kinetic energy factor ($\alpha > 1$) expresses the ratio of the true kinetic energy of the velocity profile to that of the mean flow velocity.

For laminar flow in a wide rectangular channel, $\alpha = 1.54$, and for laminar flow in a pipe, $\alpha = 2.0$. For turbulent flow in a duct $\alpha \approx 1$.

Heat transfer Q may often be ignored. The change of mechanical energy into internal energy Δu may be expressed as E_L. Flow analysis involves the change in the Bernoulli constant ($\Delta B = B_2 - B_1$) between stations 1 and 2 along the conduit, and the Bernoulli equation can be expressed as

$$\left(\frac{p}{\rho} + \alpha \frac{V^2}{2} + gz\right)_1 + E_M = \left(\frac{p}{\rho} + \alpha \frac{V^2}{2} + gz\right)_2 + E_L \tag{6a}$$

or, dividing by g, in the form as

$$\left(\frac{p}{\rho g} + \alpha \frac{V^2}{2g} + z\right)_1 + H_M = \left(\frac{p}{\rho g} + \alpha \frac{V^2}{2g} + z\right)_2 + H_L \tag{6b}$$

The factors E_M and E_L are defined as positive, where $gH_M = E_M$ represents energy added to the conduit flow by pumps or blowers, and $gH_L = E_L$ represents energy dissipated, that is, converted into heat as mechanically nonrecoverable energy. A turbine or fluid motor thus has a negative H_M or E_M. For conduit systems with branches involving inflow or outflow, the total energies must be treated, and analysis is in terms of $\dot{m}B$ and not B.

When real-fluid effects of viscosity or turbulence are included, the continuity relation in Equation (2b) is not changed, but V must be evaluated from the integral of the velocity profile, using timeaveraged local velocities.

In fluid flow past fixed boundaries, the velocity at the boundary is zero and shear stresses are produced. The equations of motion then become complex and exact solutions are difficult to find, except in simple cases.

Laminar Flow

For steady, fully developed laminar flow in a parallel-walled conduit, the shear stress τ varies linearly with distance y from the centerline. For a wide rectangular channel,

$$\tau = \left(\frac{y}{b}\right)\tau_w = \left(\mu \frac{dv}{dy}\right)$$

where

$$\tau_w = \text{wall shear stress} = b(dp/ds)$$
$$2b = \text{wall spacing}$$
$$s = \text{flow direction}$$

Because the velocity is zero at the wall ($y = b$), the integrated result is

$$v = \left(\frac{b^2 - y^2}{2\mu}\right)\frac{dp}{ds}$$

This is the **Poiseuille-flow parabolic velocity profile** for a wide rectangular channel. The average velocity V is two-thirds the maximum velocity (at $y = 0$), and the longitudinal pressure drop in terms of conduit flow velocity is

$$\frac{dp}{ds} = -\left(\frac{3\mu V}{b^2}\right) \tag{7}$$

The parabolic velocity profile can also be derived for the axisymmetric conduit (pipe) of radius R but with a different constant. The average velocity is then half the maximum, and the pressure drop relation is

$$\frac{dp}{ds} = -\left(\frac{8\mu V}{R^2}\right) \tag{8}$$

Turbulence

Fluid flows are generally turbulent, involving random perturbations or fluctuations of the flow (velocity and pressure), characterized by an extensive hierarchy of scales or frequencies (Robertson 1963). Flow disturbances that are not random, but have some degree of periodicity, such as the oscillating vortex trail behind bodies, have been erroneously identified as turbulence. Only flows involving random perturbations without any order or periodicity are turbulent; the velocity in such a flow varies with time or locale of measurement (Figure 2).

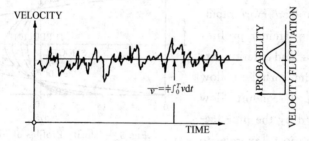

Fig. 2　Velocity Fluctuation at Point in Turbulent Flow

Turbulence can be quantified by statistical factors. Thus, the velocity most often used in velocity profiles is the temporal average velocity v, and the strength of the turbulence is characterized by the root-mean-square of the instantaneous variation in velocity about this mean. The effects of turbulence cause the fluid to diffuse momentum, heat, and mass very rapidly across the flow.

The **Reynolds number** Re, a dimensionless quantity, gives the relative ratio of inertial to viscous forces:

$$Re = VL/\nu$$

where

  L = characteristic length

  ν = kinematic viscosity

In flow through round pipes and tubes, the characteristic length is the diameter D. Generally, laminar flow in pipes can be expected if the Reynolds number, which is based on the pipe diameter, is less than about 2300. Fully turbulent flow exists when Re_D

>10000. Between 2300 and 10000, the flow is in a transition state and predictions are unreliable. In other geometries, different criteria for the Reynolds number exist.

BASIC FLOW PROCESSES

Wall Friction

At the boundary of real-fluid flow, the relative tangential velocity at the fluid surface is zero. Sometimes in turbulent flow studies, velocity at the wall may appear finite, implying a fluid slip at the wall. However, this is not the case; the difficulty is in velocity measurement (Goldstein 1938). Zero wall velocity leads to a high shear stress near the wall boundary and a slowing down of adjacent fluid layers. A velocity profile develops near a wall, with the velocity increasing from zero at the wall to an exterior value within a finite lateral distance.

Laminar and turbulent flow differ significantly in their velocity profiles. Turbulent flow profiles are flat compared to the more pointed profiles of laminar flow (Figure 3). Near the wall, velocities of the turbulent profile must drop to zero more rapidly than those of the laminar profile, so the shear stress and friction are much greater in the turbulent flow case. Fully developed conduit flow may be characterized by the **pipe factor**, which is the ratio of average to maximum (centerline) veloci-

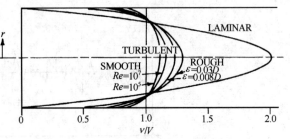

Fig. 3 Velocity Profiles of Flow in Pipes

ty. Viscous velocity profiles result in pipe factors of 0.667 and 0.50 for wide rectangular and axisymmetric conduits. Figure 4 indicates much higher values for rectangular and circular conduits for turbulent flow. Due to the flat velocity profiles, the kinetic energy factor α in Equation (6) ranges from 1.01 to 1.10 for fully developed turbulent pipe flow.

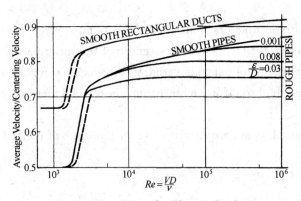

Fig. 4 Pipe Factor for Flow in Conduits

Boundary Layer

In most flows, the friction of a bounding wall on the fluid flow is evidenced by a boundary layer. For flow around bodies, this layer (which is quite thin relative to distances in the flow direction) encompasses all viscous or turbulent actions, causing the velocity in it to vary rapidly from zero at the wall to that of the outer flow at its edge. Boundary layers are generally laminar near the start of their formation but may become turbulent downstream of the transition point (Figure 5). For conduit flows, spacing between adjacent walls is generally small compared with distances in the flow direction. As a result, layers from the walls meet at the centerline to fill the conduit.

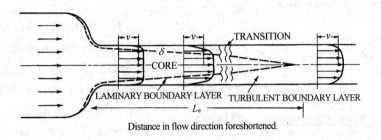

Fig. 5　Flow in Conduit Entrance Region

A significant boundary-layer occurrence exists in a pipeline or conduit following a well-rounded entrance (Figure 5). Layers grow from the walls until they meet at the center of the pipe. Near the start of the straight conduit, the layer is very thin (and laminar in all probability), so the uniform velocity core outside has a velocity only slightly greater than the average velocity. As the layer grows in thickness, the slower velocity near the wall requires a velocity increase in the uniform core to satisfy continuity. As the flow proceeds, the wall layers grow (and the centerline velocity increases) until they join, after an entrance length L_e. Application of the Bernoulli relation of Equation (5) to the core flow indicates a decrease in pressure along the layer. Ross (1956) shows that although the entrance length L_e is many diameters, the length in which the pressure drop significantly exceeds those for fully developed flow is on the order of 10 diameters for turbulent flow in smooth pipes.

In more general boundary-layer flows, as with wall layer development in a diffuser or for the layer developing along the surface of a strut or turning vane, pressure gradient effects can be severe and may even lead to separation. The development of a layer in an adverse-pressure gradient situation (velocity v_1 at edge $y=\delta$ of layer decreasing in flow direction) with separation is shown in Figure 6. Downstream from the separation point, fluid backflows near the wall. Separation is due to frictional velocity (thus local kinetic energy) reduction near the wall. Flow near the wall

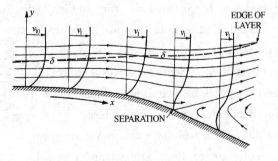

Fig. 6　Boundary Layer Flow to Separation

no longer has energy to move into the higher pressure imposed by the decrease in v_1 at the edge of the layer. The locale of this separation is difficult to predict, especially for the turbulent boundary layer. <u>Analyses verify the experimental observation that a turbulent boundary layer is less subject to separation than a laminar one because of its greater kinetic energy.</u>

Flow Patterns with Separation

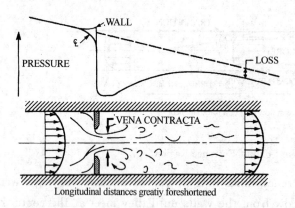

Fig. 7 Geometric Separation, Flow Development, and Loss in Flow Through Orifice

In technical applications, flow with separation is common and often accepted if it is too expensive to avoid. Flow separation may be geometric or dynamic. Dynamic separation is shown in Figure 6. Geometric separation (Figures 7 and 8) results when a fluid stream passes over a very sharp corner, as with an orifice; the fluid generally leaves the corner irrespective of how much its velocity has been reduced by friction.

For geometric separation in orifice flow (Figure 7), the outer streamlines separate from the sharp corners and, because of fluid inertia, contract to a section smaller than the orifice opening, the **vena contracta**, with a limiting area of about six-tenths of the orifice opening. After the vena contracta, the fluid stream expands rather slowly through turbulent or laminar interaction with the fluid along its sides. Outside the jet, fluid velocity is small compared to that in the jet. Turbulence helps spread out the jet, increases the losses, and brings the velocity distribution back to a more uniform profile. Finally, at a considerable distance downstream, the velocity profile returns to the fully developed flow of Figure 3.

Other geometric separations (Figure 8) occur at a sharp entrance to a conduit, at an inclined plate or damper in a conduit, and at a sudden expansion. For these, a vena contracta can be identified; for sudden expansion, its area is that of the upstream contraction. Ideal-fluid theory, using free streamlines, provides insight and predicts contraction coefficients for valves, orifices, and vanes (Robertson 1965). These geometric flow separations are large loss-producing devices. To expand a flow efficiently or to have an entrance with minimum losses, the device should be

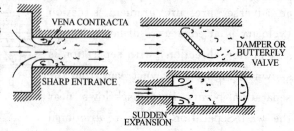

Fig. 8 Examples of Geometric Separation Encountered in Flows in Conduits

designed with gradual contours, a diffuser, or a rounded entrance.

Flow devices with gradual contours are subject to separation that is more difficult to predict, because it involves the dynamics of boundary layer growth under an adverse pressure gradient rather than flow over a sharp corner. In a diffuser, which is used to reduce the loss in expansion, it is possible to expand the fluid some distance at a gentle angle without difficulty (particularly if the boundary layer is turbulent). Eventually, separation may occur (Figure 9), which is frequently asymmetrical because of irregularities. Downstream flow involves flow reversal (backflow) and excess losses exist. Such separation is termed **stall** (Kline 1959). Larger area expansions may use splitters that divide the diffuser into smaller divisions less likely to have separations (Moore and Kline 1958). Another technique for controlling separation is to bleed some low-velocity fluid near the wall (Furuya et al. 1976). Alternatively, Heskested (1965, 1970) shows that suction at the corner of a sudden expansion has a strong positive effect on geometric separation.

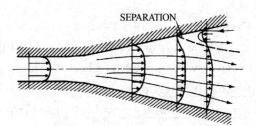

Fig. 9 Separation in Flow in Diffuser

Drag Forces on Bodies or Struts

Bodies in moving fluid streams are subjected to appreciable fluid forces or drag. Conventionally expressed in coefficient form, drag forces on bodies can be expressed as

$$D = C_D \rho A V^2 / 2 \qquad (9)$$

where A is the projected (normal to flow) area of the body. The **drag coefficient** C_D depends on the body's shape and angularity and on the Reynolds number of the relative flow in terms of the body's characteristic dimension.

For Reynolds numbers of 10^3 to above 10^5, the C_D of most bodies is constant due to flow separation, but above 10^5, the C_D of rounded bodies drops suddenly as the surface boundary layer undergoes transition to turbulence. Typical C_D values are given in Table 1; Hoerner (1965) gives expanded values.

Drag Coefficients — Table 1

Body Shape	$10^3 < Re < 2 \times 10^5$	$Re > 3 \times 10^5$	Body Shape	$10^3 < Re < 2 \times 10^5$	$Re > 3 \times 10^5$
Sphere	0.36 to 0.47	~0.1	Circular cylinder	1.0 to 1.1	0.35
Disk	1.12	1.12	Elongated rectangular strut	1.0 to 1.2	1.0 to 1.2
Streamlined strut	0.1 to 0.3	<0.1	Square strut	~2.0	~2.0

For a strut crossing a conduit, the contribution to the loss of Equation (6b) is

$$H_L = C_D \left(\frac{A}{A_c}\right)\left(\frac{V^2}{2g}\right) \qquad (10)$$

where

A_c = conduit cross-sectional area

A = area of the strut facing the flow

Cavitation

Liquid flow with gas- or vapor-filled pockets can occur if the absolute pressure is reduced to vapor pressure or less. In this case, a cavity or series of cavities forms, because liquids are rarely pure enough to withstand any tensile stressing or pressures less than vapor pressure for any length of time (John and Haberman 1980, Knapp et al. 1970, Robertson and Wislicenus 1969). Robertson and Wislicenus (1969) indicate significant occurrences in various technical fields, chiefly in hydraulic equipment and turbomachines.

Initial evidence of cavitation is the collapse noise of many small bubbles that appear initially as they are carried by the flow into regions of higher pressure. The noise is not deleterious and serves as a warning of the occurrence. As flow velocity further increases or pressure decreases, the severity of cavitation increases. More bubbles appear and may join to form large fixed cavities. The space they occupy becomes large enough to modify the flow pattern and alter performance of the flow device. Collapse of the cavities on or near solid boundaries becomes so frequent that the cumulative impact in time results in damage in the form of cavitational erosion of the surface or excessive vibration. As a result, pumps can lose efficiency or their parts may erode locally. Control valves may be noisy or seriously damaged by cavitation.

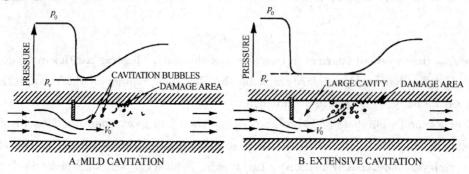

Fig. 10 Cavitation in Flows in Orifice or Valve

Cavitation in orifice and valve flow is indicated in Figure 10. With high upstream pressure and a low flow rate, no cavitation occurs. As pressure is reduced or flow rate increased, the minimum pressure in the flow (in the shear layer leaving the edge of the orifice) eventually approaches vapor pressure. Turbulence in this layer causes fluctuating pressures below the mean (as in vortex cores) and small bubble-like cavities. These are carried downstream into the region of pressure regain where they collapse, either in the fluid or on the wall (Figure 10A). As the pressure is reduced, more vapor- or gas-filled bubbles result and coalesce into larger ones. Eventually, a single large cavity results that collapses further downstream (Figure 10B). The region of wall damage is then as many as 20 diameters downstream from the valve or orifice plate.

Sensitivity of a device to cavitation occurrence is measured by the **cavitation index** or **cavitation number**, which is the ratio of the available pressure above vapor pressure to the dynamic pressure of the reference flow:

$$\sigma = \frac{2(p_0 - p_v)}{\rho V_0^2} \tag{11}$$

where p_v is the vapor pressure, and the subscript o refers to appropriate reference conditions. Valve analyses use such an index in order to determine when cavitation will affect the discharge coefficient (Ball 1957). With flow-metering devices such as orifices, venturis, and flow nozzles, there is little cavitation, because it occurs mostly downstream of the flow regions involved in establishing the metering action.

The detrimental effects of cavitation can be avoided by operating the liquid-flow device at high enough pressures. When this is not possible, the flow must be changed or the device must be built to withstand cavitation effects. Some materials or surface coatings are more resistant to cavitation erosion than others, but none is immune. Surface contours can be designed to delay the onset of cavitation.

Nonisothermal Effects

When appreciable temperature variations exist, the primary fluid properties (density and viscosity) are no longer constant, as usually assumed, but vary across or along the flow. The Bernoulli equation in the form of Equations (5a) through (5c) must be used, because volumetric flow is not constant. With gas flows, the thermodynamic process involved must be considered. In general, this is assessed in applying Equation (5a), written in the following form:

$$\int \frac{dp}{\rho} + \frac{V^2}{2} + gz = B \tag{12}$$

Effects of viscosity variations also appear. With nonisothermal laminar flow, the parabolic velocity profile (Figure 3) is no longer valid. For gases, viscosity increases as the square root of absolute temperature, and for liquids, it decreases with increasing temperature. This results in opposite effects.

For fully developed pipe flow, the linear variation in shear stress from the wall value τ_w to zero at the centerline is independent of the temperature gradient. In the section on Laminar Flow, τ is defined as $\tau = (y/b)\tau_w$, where y is the distance from the centerline and $2b$ is the wall spacing. For pipe radius $R = D/2$ and distance from the wall $y = R - r$ (see Figure 11), then $\tau = \tau_w(R-y)/R$. Then, solving Equation (1) for the change in veloci-

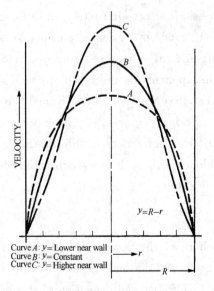

Fig. 11 Effect of Viscosity Variation on Velocity Profile of Laminar Flow in Pipe

part I Theory

ty gives

$$dv = \left[\frac{\tau_w(R-y)}{R\mu}\right]dy = -\left(\frac{\tau_w}{R\mu}\right)r\,dr \qquad (13)$$

When the fluid has a lower viscosity near the wall than at the center (due to external heating of liquid or cooling of gas via heat transfer through the pipe wall), the velocity gradient is steeper near the wall and flatter near the center, so the profile is generally flattened. When liquid is cooled or gas is heated, the velocity profile becomes more pointed for laminar flow (Figure 11). Calculations were made for such flows of gases and liquid metals in pipes (Deissler 1951). Occurrences in turbulent flow are less apparent. If enough heating is applied to gaseous flows, the viscosity increase can cause reversion to laminar flow.

Buoyancy effects and gradual approach of the fluid temperature to equilibrium with that outside the pipe can cause considerable variation in the velocity profile along the conduit. Thus, Colborne and Drobitch (1966) found the pipe factor for upward vertical flow of hot air at a Reynolds number less than 2000 reduced to about 0.6 at 40 diameters from the entrance, then increased to about 0.8 at 210 diameters, and finally decreased to the isothermal value of 0.5 at the end of 320 diameters.

Compressibility

All fluids are compressible to some degree; their density depends on the pressure. Steady liquid flow may ordinarily be treated as incompressible, and incompressible flow analysis is satisfactory for gases and vapors at velocities below about 20 to 40m/s, except in long conduits.

For liquids in pipelines, if flow is suddenly stopped, a severe pressure surge or water hammer is produced that travels along the pipe at the speed of sound in the liquid. This pressure surge alternately compresses and decompresses the liquid. For steady gas flows in long conduits, a decrease in pressure along the conduit can reduce the density of the gas significantly enough to cause the velocity to increase. If the conduit is long enough, velocities approaching the speed of sound are possible at the discharge end, and the Mach number (the ratio of the flow velocity to the speed of sound) must be considered.

Some compressible flows occur without heat gain or loss (adiabatically). If there is no friction (conversion of flow mechanical energy into internal energy), the process is reversible as well. Such a reversible adiabatic process is called isentropic, and follows the relationship

$$p/\rho^k = \text{constant}$$
$$k = c_p/c_v$$

where k, the ratio of specific heats at constant pressure and volume, has a value of 1.4 for air and diatomic gases.

The Bernoulli equation of steady flow, Equation (12), as an integral of the ideal-fluid equation of motion along a streamline, then becomes

$$\int \frac{dp}{\rho} + \frac{V^2}{2} = \text{constant} \tag{14}$$

where, as in most compressible flow analyses, the elevation terms involving z are insignificant and are dropped.

For a frictionless adiabatic process, the pressure term has the form

$$\int_1^2 \frac{dp}{\rho} = \frac{k}{k-1}\left(\frac{p_2}{\rho_2} - \frac{p_1}{\rho_1}\right) \tag{15}$$

Then, between stations 1 and 2 for the isentropic process,

$$\frac{p_1}{\rho_1}\left(\frac{k}{k-1}\right)\left[\left(\frac{p_2}{p_1}\right)^{(k-1)/k} - 1\right] + \frac{V_2^2 - V_1^2}{2} = 0 \tag{16}$$

Equation (16) replaces the Bernoulli equation for compressible flows and may be applied to the stagnation point at the front of a body. With this point as station 2 and the upstream reference flow ahead of the influence of the body as station 1, $V_2 = 0$. Solving Equation (16) for p_2 gives

$$p_s = p_2 = p_1\left[1 + \left(\frac{k-1}{2}\right)\frac{\rho_1 V_1^2}{k p_1}\right]^{k/(k-1)} \tag{17}$$

where p_s is the stagnation pressure.

Because kp/ρ is the square of the acoustic velocity a and the Mach number $M = V/a$, the stagnation pressure relation becomes

$$p_s = p_1\left[1 + \left(\frac{k-1}{2}\right)M_1^2\right]^{k/(k-1)} \tag{18}$$

For Mach numbers less than one,

$$p_s = p_1 + \frac{\rho_1 V_1^2}{2}\left[1 + \frac{M_1^2}{4} + \left(\frac{2-k}{24}\right)M_1^4 + \cdots\right] \tag{19}$$

When $M = 0$, Equation (19) reduces to the incompressible flow result obtained from Equation (5a). Appreciable differences appear when the Mach number of the approaching flow exceeds 0.2. Thus a pitot tube in air is influenced by compressibility at velocities over about 66m/s.

Flows through a converging conduit, as in a flow nozzle, venturi, or orifice meter, also may be considered isentropic. Velocity at the upstream station 1 is negligible. From Equation (16), velocity at the downstream station is

$$V_2 = \sqrt{\frac{2k}{k-1}\left(\frac{p_1}{\rho_1}\right)\left[1 - \left(\frac{p_2}{p_1}\right)^{(k-1)/k}\right]} \tag{20}$$

The mass flow rate is

$$\dot{m} = V_2 A_2 \rho_2 = A_2\sqrt{\frac{2k}{k-1}(p_1 \rho_1)\left[\left(\frac{p_2}{p_1}\right)^{2/k} - \left(\frac{p_2}{p_1}\right)^{(k+1)/k}\right]} \tag{21}$$

The corresponding incompressible flow relation is

$$\dot{m}_{in} = A_2 \rho\sqrt{2\Delta p/\rho} = A_2\sqrt{2\rho(p_1 - p_2)} \tag{22}$$

The compressibility effect is often accounted for in the **expansion factor Y**:

$$\dot{m} = Y \dot{m}_{in} = A_2 Y\sqrt{2\rho(p_1 - p_2)} \tag{23}$$

Y is 1.00 for the incompressible case. For air ($k=1.4$), a Y value of 0.95 is reached with orifices at $p_2/p_1=0.83$ and with venturis at about 0.90, when these devices are of relatively small diameter (D_2/D_1 less than 0.5).

As p_2/p_1 decreases, the flow rate increases, but more slowly than for the incompressible case because of the nearly linear decrease in Y. However, the downstream velocity reaches the local acoustic value and the discharge levels off at a value fixed by upstream pressure and density at the critical ratio:

$$\left.\frac{p_2}{p_1}\right|_c = \left(\frac{2}{k+1}\right)^{k/(k-1)} = 0.53 \text{ for air} \tag{24}$$

At higher pressure ratios than critical, **choking** (no increase in flow with decrease in downstream pressure) occurs and is used in some flow control devices to avoid flow dependence on downstream conditions.

WORDS AND EXPRESSIONS

acoustic	*adj.* 有关声音的，声学的，音响学的
adiabatic	*adj.* [物] 绝热的，隔热的
adverse	*adj.* 不利的，敌对的，相反的
alternatively	*adv.* 做为选择，二者择一地
approximation	*n.* 接近，走近，[数] 近似值
asymmetrical	*adj.* 不均匀的，不对称的
axisymmetric	*adj.* 轴对称的
buoyancy	*n.* 浮性，浮力，轻快
cavitation	*n.* 气穴现象
choking	*adj.* 窒息的，憋闷的，透不过气来的
coalesce	*v.* 接合
conduit	*n.* 管道，导管，沟渠，泉水，喷泉
conservation	*n.* 保存，保持，守恒
criteria	*n.* 标准
cross section	横截面
cumulative	*adj.* 累积的
decompress	*v.* (使)减压 (compress *v.* 压缩，加压)
deform	*v.* (使)变形
deleterious	*adj.* 有害的，有毒的
diatomic	*adj.* 二原子的，二氢氧基的，二价的
differential equation	微分方程
diffuser	*n.* 扩散段；扩散管；扩散器；扩散体；扩压器；扬声器纸盆；散光罩；进气道；浸射体；偏光片；漫射器；漫射体；散射器；散射体
discrepancy	*n.* 相差，差异，矛盾

discrete	*adj.* 不连续的，离散的
displacement	*n.* 移置，转移，取代，置换，位移，排水量
duct	*n.* 管，输送管，排泄管；*vt.* 通过管道输送
elevation	*n.* 上升，高地，立面图，海拔，提高，仰角，崇高，壮严
encompass	*v.* 包围，环绕，包含或包括某事物
fitting	*adj.* 适合的，相称的，适宜的；*n.* 试穿，试衣，装配，装置
flanged	*adj.* 带凸缘的，用法兰连接的，折边的，翼缘的
fluctuation	*n.* 波动，起伏
frequency	*n.* 频率，周率，发生次数
gradient	*adj.* 倾斜的；*n.* 梯度，倾斜度，坡度
hammer	*n.* 铁锤，槌，锤子；*v.* 锤击，锤打
harmonics	*n.* 和声学
hierarchy	*n.* 层次，层级
homogeneous	*adj.* 同类的，相似的，均一的，均匀的
hydrostatic	*adj.* 静水力学的，流体静力学的
identical	*adj.* 同一的，同样的
illustrated	*n.* 有插画的报章杂志；*adj.* 有插图的
immune	*adj.* 免疫的
in accord with	符合，一致
in the nature of	具有...性质
inertia	*n.* 惯性，惯量
integrated	*adj.* 综合的，完整的，集成的
isentropic	*adj.* ［物］等熵的
kinetic	*adj.* （运）动的，动力（学）的
lateral	*n.* 侧部，支线，边音；*adj.* 横（向）的，侧面的
longitudinal	*adj.* 经度的，纵向的
magnitude	*n.* 大小，数量，巨大，广大，量级
mechanics	*n.* （用作单数）机械学、力学，（用作复数）技巧，结构
metering	测量(法)，计［配］量，测定
Newtonian	*adj.* 牛顿的，牛顿学说的；*n.* 信仰牛顿学说的人
nonisothermal	*adj.* 非等温的
normal to	垂直于
nozzle	*n.* 管口，喷嘴
obstruction	*n.* 阻塞，妨碍，障碍物
onset	*n.* 攻击，进攻，有力的开始，肇端；*n.* ［医］发作
orifice	*n.* 孔，口
oscillate	*v.* 振荡
perforate	*v.* 打孔
periodicity	*n.* 周期

part I Theory

pertinent	*adj.* 有关的，相干的，中肯的
Pitot tube	皮托管
profile	*n.* 剖面，侧面，外形，轮廓，分布图
proportional	*adj.* 比例的，成比例的，相称的，均衡的
quantify	*vt.* 确定数量；*v.* 量化
reinforce	*vt.* 加强，增援，补充，增加…的数量，修补，加固
	vi. 求援，得到增援；*n.* 加固物
reservoir	*n.* 水库，蓄水池
shear	*v.* 剪，修剪，剪切；*n.* 剪力
splitter	分离机；分离器；分离设备；分离柱；分裂器；
	分裂设备；分流片；分流器；导流板
statistical	*adj.* 统计的，统计学的
steep	*adj.* 陡峭的，险峻的，急剧升降的，不合理的；*n.* 悬崖，峭壁，浸渍，浸渍液；*v.* 浸，泡，沉浸
strut	*n.* 高视阔步，支柱，压杆
surge	*n.* 巨涌，汹涌，澎湃，喘振；*vi.* 汹涌，澎湃，振荡，滑脱，放松 *vt.* 使汹涌奔腾，急放
tangential	*adj.* 切线的
temporal	*adj.* 时间的，当时的，暂时的，现世的，世俗的，[解]颞的
tensile	*adj.* 可拉长的，可伸长的，[物]张力的，拉力的
thereof	*adv.* 在其中，关于…，将它，它的
thermodynamics	*n.* [物]热力学
turbulence	*n.* 骚乱，动荡，(液体或气体的)紊乱，湍流
upstream	*adv.* 向上游，溯流，逆流地；*adj.* 溯流而上的，上游的(downstream；*adv.* 下游地；*adj.* 下游的)
valve	*n.* 阀，[英]电子管，真空管
vena	*n.* <拉>静脉
vena contracta	射流紧缩，缩脉
vibration	*n.* 振动，颤动，摇动，摆动
viscosity	*n.* 黏质，黏性
viscous	*adj.* 黏性的，黏滞的，胶粘的
vortex	*n.* 旋涡，旋风，涡流，(动乱，争论等的)中心
water hammer	水击，水击作用
withstand	*vt.* 抵挡，经受住

NOTATIONS

(1) The velocity gradient associated with viscous shear for a simple case <u>involving</u> flow velocity in the x direction but of varying magnitude in the y direction is illustrated in Figure 1B.

图 1B 中图示出一个和黏性剪力有关的速度梯度简单例子，包括 x 方向上的流动速度以及 y 方向上流动速度的变化大小。

(2) Both ρ and V may vary over the cross section A of the conduit. If both ρ and V are constant over the cross-sectional area normal to the flow, then

$$\dot{m} = \rho V A = \text{constant}$$

where $\dot{m}$ is the mass flow rate across the area normal to the flow.

ρ 和 V 都可能在管道的横截面 A 上变化。如果在垂直于流动的横截面上，ρ 和 V 都是常数，那么 $\dot{m} = \rho V A = $ 常数，其中 $\dot{m}$ 是通过垂直于流动的面的质量流率。

(3) In cases with no viscous action and no work interaction, B is constant; more generally its change (or lack thereof) is considered in applying the Bernoulli equation.

在没有黏性作用和功的相互作用的情况下，B 是常数；更为一般的是在应用伯努利方程时考虑（或者不考虑）它的变化。

(4) Analyses verify the experimental observation that a turbulent boundary layer is less subject to separation than a laminar one because of its greater kinetic energy.

分析说明，因为较大的紊流动能，紊流边界层比层流边界层更不易分离的实验观察是正确的。

(5) In this case, a cavity or series of cavities forms, because liquids are rarely pure enough to withstand any tensile stressing or pressures less than vapor pressure for any length of time.

在这种情况下，因为液体不足够纯净，以至不能承受任一时间长度的小于气体压力的张力或压力，形成了一个或一连串的气穴。

(6) Pay attention to the following phrases（注意下列短语用法）：

Liquid viscosity <u>decreases with increasing</u>（随……增大而减小） temperature.

Fluid energy <u>is composed of</u>（由……组成） kinetic, potential (due to elevation Z), and internal (u) energies.

Rearranging, the energy equation <u>can be written as</u>（可被写成） the generalized Bernoulli equation.

Analyses verify the experimental observation that a turbulent boundary layer <u>is</u> less <u>subject to</u>（受支配，从属于，可以…的，易遭受…） separation than a laminar one because of its greater kinetic energy.

EXERCISES

请翻译下列句子。

1) Fluids differ from solids in their reaction to shearing. When placed under shear stress, a solid deforms only a finite amount, whereas a fluid deforms continuously for as long as the shear is applied. Both liquids and gases are fluids. Although liquids and gases differ strongly in the nature of molecular actions, their primary mechanical differences are in the degree of compressibility and liquid formation of a free surface (interface).

2) This section considers homogeneous, constant-property, incompressible fluids and introduces fluid dynamic considerations used in most analyses.

3) For the ideal-fluid model, flow patterns around bodies (or in conduit section changes) result from displacement effects. An obstruction in a fluid stream, such as a strut in a flow or a bump on the conduit wall, pushes the flow smoothly out of the way, so that behind the obstruction, the flow becomes uniform again. The effect of fluid inertia (density) appears only in pressure changes.

4) For steady, fully developed laminar flow in a parallel-walled conduit, the shear stress τ varies linearly with distance y from the centerline.

5) In more general boundary-layer flows, as with wall layer development in a diffuser or for the layer developing along the surface of a strut or turning vane, pressure gradient effects can be severe and may even lead to separation.

6) Flows through a converging conduit, as in a flow nozzle, venturi, or orifice meter, also may be considered isentropic.

LESSON 3

HEAT TRANSFER

HEAT is energy in transit due to a temperature difference. The thermal energy is transferred from one region to another by three modes of **heat transfer**: conduction, convection, and radiation. Heat transfer is among a group of energy transport phenomena that includes mass transfer, momentum transfer or fluid friction and electrical conduction. Transport phenomena have similar rate equations, in which flux is proportional to a potential difference. In heat transfer by conduction and convection, the potential difference is the temperature difference. Heat, mass, and momentum transfer are often considered together because of their similarities and interrelationship in many common physical processes.

This chapter presents the elementary principles of single-phase heat transfer with emphasis on heating, refrigerating, and air conditioning.

HEAT TRANSFER PROCESSES

Thermal Conduction. This is the mechanism of heat transfer whereby energy is transported between parts of a continuum by the transfer of kinetic energy between particles or groups of particles at the atomic level. In gases, conduction is caused by elastic collision of molecules; in liquids and electrically nonconducting solids, it is believed to be caused by longitudinal oscillations of the lattice structure. Thermal conduction in metals occurs, like electrical conduction, through the motion of free electrons. Thermal energy transfer occurs in the direction of decreasing temperature, a consequence of the second law of thermodynamics. In solid opaque bodies, thermal conduction is the significant heat transfer mechanism because no net material flows in the process and radiation is not a factor. With flowing fluids, thermal conduction dominates in the region very close to a solid boundary, where the flow is **laminar** and parallel to the surface and where there is no eddy motion.

Thermal Convection. This form of heat transfer involves energy transfer by fluid movement and molecular conduction (Burmeister 1983, Kays and Crawford 1980). Consider heat transfer to a fluid flowing inside a pipe. If the Reynolds number is large enough, three different flow regions exist. Immediately adjacent to the wall is a **laminar sublayer** where heat transfer occurs by thermal conduction; outside the laminar sublayer is

---

Extracted from Chapter 3 of the ASHREA *Handbook-Fundamentals*.

a transition region called the **buffer layer**, where both eddy mixing and conduction effects are significant; beyond the buffer layer and extending to the center of the pipe is the **turbulent region**, where the dominant mechanism of transfer is eddy mixing.

In most equipment, the main body of fluid is in turbulent flow, and the laminar layer exists at the solid walls only. In cases of low-velocity flow in small tubes, or with viscous liquids such as glycol (i.e., at low Reynolds numbers), the entire flow may be laminar with no transition or turbulent region.

When fluid currents are produced by external sources (for example, a blower or pump), the solid-to-fluid heat transfer is termed **forced convection**. If the fluid flow is generated internally by nonhomogeneous densities caused by temperature variation, the heat transfer is termed **free convection** or **natural convection**.

Thermal Radiation. In conduction and convection, heat transfer takes place through matter. In thermal radiation, there is a change in energy form from internal energy at the source to electromagnetic energy for transmission, then back to internal energy at the receiver. Whereas conduction and convection heat transfer rates are driven primarily by temperature difference and somewhat by temperature level, radiative heat transfer rates increase rapidly with temperature levels (for the same temperature difference).

Although some generalized heat transfer equations have been mathematically derived from fundamentals, they are usually obtained from correlations of experimental data. Normally, the correlations employ certain dimensionless numbers, shown in Table 1, that are derived from dimensional analysis or analogy.

Dimensionless Numbers Commonly Used in Heat Transfer   Table 1

Name	Symbol	Value[a]	Application
Nusselt number	Nu	hD/k, hL/k, $q''D/\Delta tk$, or $q''L/\Delta tk$	Natural or forced convection, boiling or condensing
Reynolds number	Re	GD/μ or $\rho VL/\mu$	Forced convection, boiling or condensing
Prandtl number	Pr	$\mu c_p/k$	Natural or forced convection, boiling or condensing
Stanton number	St	h/Gc_p	Forced convection
Grashof number	Gr	$L^3\rho^2\beta g\Delta t/\mu^2$ or $L^3\rho^2 g\Delta t/T\mu^2$	Natural convection (for ideal gases)
Fourier number	Fo	$\alpha\tau/L^2$	Unsteady-state conduction
Peclet number	Pe	GDc_p/k or Re Pr	Forced convection (small Pr)
Graetz number	Gz	GD^2c_p/kL or Re Pr D/L	Laminar forced convection

[a] A list of the other symbols used in this chapter appears in the section on Symbols.

STEADY-STATE CONDUCTION

For steady-state heat conduction in one dimension, the Fourier law is

$$q = -(kA)\frac{dt}{dx} \tag{1}$$

where

q = heat flow rate, W

k = thermal conductivity, W/(m · K)

A = cross-sectional area normal to flow, m²

dt/dx = temperature gradient, K/m

Equation (1) states that the heat flow rate q in the x direction is directly proportional to the temperature gradient dt/dx and the cross-sectional area A normal to the heat flow. The proportionality factor is the thermal conductivity k. The minus sign indicates that the heat flow is positive in the direction of decreasing temperature. Conductivity values are sometimes given in other units, but consistent units must be used in Equation (1).

Equation (1) may be integrated along a path of constant heat flow rate to obtain

$$q = k\left(\frac{A_m}{L_m}\right)\Delta t = \frac{\Delta t}{R} \qquad (2)$$

where

A_m = mean cross-sectional area normal to flow, m²

L_m = mean length of heat flow path, m

Δt = overall temperature difference, K

R = thermal resistance, K/W

Thermal resistance R is directly proportional to the mean length L_m of the heat flow path and inversely proportional to the conductivity k and the mean cross-sectional area A_m normal to the flow. Equations for thermal resistances of a few common shapes are given in Table 2. Mathematical solutions to many heat conduction problems are addressed by Carslaw and Jaeger (1959). Complicated problems can be solved by graphical or numerical methods such as described by Croft and Lilley (1977), Adams and Rogers (1973), and Patankar (1980).

Solutions for Some Steady-State Thermal Conduction Problems   Table 2

System	R in Equation $q=\Delta t/R$	System	R in Equation $q=\Delta t/R$
Flat wall or curved wall if curvature is small (wall thickness less than 0.1 of inside diameter) SURFACE AREA, A	$R = \dfrac{L}{kA}$	Buried cylinder LONG CYLINDER OF LENGTH, L $\Delta t = t_s - t_p$	$R = \dfrac{\ln\left[(a+\sqrt{a^2-r^2})/r\right]}{2\pi kL}$ $= \dfrac{\cosh^{-1}(a/r)}{2\pi kL}$ $(L \gg 2r)$
Radial flow through a right circular cylinder LONG CYLINDER OF LENGTH, L	$R = \dfrac{\ln(r_o/r_i)}{2\pi kL}$	Radial flow in a hollow sphere	$R = \dfrac{(1/r_i - 1/r_o)}{4\pi k}$

L, r, a = dimensions, m

k = thermal conductivity at average material temperature, W/(m·K)

A = surface area, m²

part Ⅰ Theory

Analogy to Electrical Conduction. Equation (2) is analogous to Ohm's law for electrical circuits: thermal current (heat flow) in a **thermal circuit** is directly proportional to the thermal potential (temperature difference) and inversely proportional to the thermal resistance. This electrical-thermal analogy can be used for heat conduction in complex shapes that resist solution by exact analytical means. The thermal circuit concept is also useful for problems involving combined conduction, convection, and radiation.

OVERALL HEAT TRANSFER

In most steady-state heat transfer problems, more than one heat transfer mode is involved. The various heat transfer coefficients may be combined into an overall coefficient so that the total heat transfer can be calculated from the terminal temperatures. The solution to this problem is much simpler if the concept of a thermal circuit is employed.

Local Overall Heat Transfer Coefficient——Resistance Method

Consider heat transfer from one fluid to another by a three-step steady-state process: from a warmer fluid to a solid wall, through the wall, then to a colder fluid. An **overall heat transfer coefficient** U based on the difference between the bulk temperatures $t_1 - t_2$ of the two fluids is defined as follows:

$$q = UA(t_1 - t_2) \tag{3}$$

where A is the surface area. Because Equation (3) is a definition of U, the surface area A on which U is based is arbitrary; it should always be specified in referring to U.

The temperature drops across each part of the heat flow path are

$$t_1 - t_{s1} = qR_1$$
$$t_{s1} - t_{s2} = qR_2$$
$$t_{s2} - t_2 = qR_3$$

where t_{s1}, and t_{s2} are the warm and cold surface temperatures of the wall, respectively, and R_1, R_2, and R_3 are the thermal resistances. Because the same quantity of heat flows through each thermal resistance, these equations combined yield the following:

$$\frac{t_1 - t_2}{q} = \frac{1}{UA} = R_1 + R_2 + R_3 \tag{4}$$

As shown above, the equations are analogous to those for electrical circuits; for thermal current flowing through n resistances in *series*, the resistances are additive:

$$R_0 = R_1 + R_2 + R_3 + \cdots + R_n \tag{5}$$

Similarly, **conductance** is the reciprocal of resistance, and for heat flow through resistances in *parallel*, the conductances are additive:

$$C = \frac{1}{R_0} = \frac{1}{R_1} + \frac{1}{R_2} + \frac{1}{R_3} + \cdots + \frac{1}{R_n} \tag{6}$$

For convection, the thermal resistance is inversely proportional to the **convection coefficient** h_c and the applicable surface area:

$$R_c = \frac{1}{h_c A} \tag{7}$$

The thermal resistance for radiation is written similarly to that for convection:

$$R_r = \frac{1}{h_r A} \tag{8}$$

The **radiation coefficient** h_r is a function of the temperatures, radiation properties, and geometrical arrangement of the enclosure and the body in question.

Resistance Method Analysis. Analysis by the resistance method can be illustrated by considering heat transfer from air outside to cold water inside an insulated pipe. The temperature gradients and the nature of the resistance analysis are shown in Figure 1.

Because air is sensibly transparent to radiation, some heat transfer occurs by both radiation and convection to the outer insula-

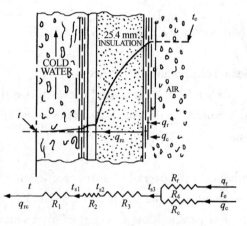

Fig. 1 Thermal Circuit Diagram for Insulated Cold Water Line

tion surface. The mechanisms act in parallel on the air side. The total transfer then passes through the insulating layer and the pipe wall by thermal conduction, and then by convection and radiation into the cold water stream. (Radiation is not significant on the water side because liquids are sensibly opaque to radiation, although water transmits energy in the visible region.) The contact resistance between the insulation and the pipe wall is assumed negligible.

The heat transfer rate q_{rc} for a given length L of pipe may be thought of as the sum of the rates q_r and q_c flowing through the parallel resistances R_r and R_c associated with the surface radiation and convection coefficients. The total flow then proceeds through the resistance R_3 offered to thermal conduction by the insulation, through the pipe wall resistance R_2, and into the water stream through the convection resistance R_1. Note the analogy to direct-current electricity. A temperature (potential) drop is required to overcome resistances to the flow of thermal current. The total resistance to heat transfer R_0 is the sum of the individual resistances:

$$R_0 = R_1 + R_2 + R_3 + R_4 \tag{9}$$

where the resultant parallel resistance R_4 is obtained from

$$\frac{1}{R_4} = \frac{1}{R_r} + \frac{1}{R_c} \tag{10}$$

If the individual resistances can be evaluated, the total resistance can be obtained from this relation. The heat transfer rate for the length of pipe L can be established by

$$q_{rc} = \frac{t_e - t}{R_0} \tag{11}$$

For a unit length of the pipe, the heat transfer rate is

$$\frac{q_{rc}}{L} = \frac{t_e - t}{R_0 L} \tag{12}$$

The temperature drop Δt through each individual resistance may then be calculated from the relation:

$$\Delta t_n = R_n q_{rc} \tag{13}$$

where $n = 1, 2,$ and 3.

Mean Temperature Difference

When heat is exchanged between two fluids flowing through a heat exchanger, the local temperature difference Δt varies along the flow path. Heat transfer may be calculated using

$$q = UA\Delta t_m \tag{14}$$

where U is the overall uniform coefficient of heat transfer from fluid to fluid, A is the area associated with the coefficient U, and Δt_m is the appropriate mean temperature difference.

For parallel flow or counterflow exchangers and for any exchanger in which one fluid temperature is substantially constant, the mean temperature difference is

$$\Delta t_m = \frac{\Delta t_1 - \Delta t_2}{\ln(\Delta t_1 / \Delta t_2)} = \frac{\Delta t_1 - \Delta t_2}{2.3 \log(\Delta t_1 / \Delta t_2)} \tag{15}$$

where Δt_1, and Δt_2 are the temperature differences between the fluids at each end of the heat exchanger. Δt_m is called the **logarithmic mean temperature difference.** For the special case of $\Delta t_1 = \Delta t_2$ (possible only with a counterflow heat exchanger with equal capacities), which leads to an indeterminate form of Equation (15), $\Delta t_m = \Delta t_1 = \Delta t_2$.

Equation (15) for Δt_m is true only if the overall coefficient and the specific heat of the fluids are constant through the heat exchanger, and no heat losses occur (often well-approximated in practice). Parker et al. (1969) give a procedure for cases with variable overall coefficient U.

Calculations using Equation (14) and Δt_m are convenient when terminal temperatures are known. In many cases, however, the temperatures of the fluids leaving the exchanger are not known. To avoid trial-and-error calculations, an alternate method involves the use of three nondimensional parameters, defined as follows:

1. Exchanger Heat Transfer Effectiveness ε

$$\varepsilon = \frac{(t_{hi} - t_{ho})}{(t_{hi} - t_{ci})} \text{ when } C_h = C_{min}$$
$$\varepsilon = \frac{(t_{co} - t_{ci})}{(t_{hi} - t_{ci})} \text{ when } C_c = C_{min} \tag{16}$$

where

$C_h = (\dot{m}c_p)_h =$ hot fluid capacity rate, W/K

$C_c = (\dot{m}c_p)_c =$ cold fluid capacity rate, W/K

$C_{min} =$ smaller of capacity rates C_h and C_c

t_h = terminal temperature of hot fluid, ℃. Subscript i indicates entering condition; subscript o indicates leaving condition.

t_c = terminal temperature of cold fluid, ℃. Subscripts i and o are the same as for t_h.

2. Number of Exchanger Heat Transfer Units (NTU)

$$\text{NTU} = \frac{AU_{avg}}{C_{min}} = \frac{1}{C_{min}} \int_A U dA \qquad (17)$$

where A is the area used to define overall coefficient U.

3. Capacity Rate Ratio Z

$$Z = \frac{C_{min}}{C_{max}} \qquad (18)$$

For a given exchanger, the heat transfer effectiveness can generally be expressed for a given exchanger as a function of the number of transfer units and the capacity rate ratio:

$$\varepsilon = f(\text{NTU}, Z, \text{flow arrangement}) \qquad (19)$$

The effectiveness is independent of the temperatures in the exchanger. For any exchanger in which the capacity rate ratio Z is zero (where one fluid undergoes a phase change; e.g., in a condenser or evaporator), the effectiveness is

$$\varepsilon = 1 - \exp(-\text{NTU}) \qquad (20)$$

Heat transferred can be determined from

$$q = C_h(t_{hi} - t_{ho}) = C_c(t_{co} - t_{ci}) \qquad (21)$$

Combining Equations (16) and (21) produces an expression for heat transfer rate in terms of entering fluid temperatures:

$$q = \varepsilon C_{min}(t_{hi} - t_{ci}) \qquad (22)$$

The proper mean temperature difference for Equation (14) is then given by

$$\Delta t_m = \frac{(t_{hi} - t_{ci})\varepsilon}{\text{NTU}} \qquad (23)$$

The effectiveness for **parallel flow exchangers** is

$$\varepsilon = \frac{1 - \exp[-\text{NTU}(1+Z)]}{1+Z} \qquad (24)$$

For $Z = 1$,

$$\varepsilon = \frac{1 - \exp(-2\,\text{NTU})}{2} \qquad (25)$$

The effectiveness for **counterflow exchangers** is

$$\varepsilon = \frac{1 - \exp[-\text{NTU}(1-Z)]}{1 - Z\exp[-\text{NTU}(1-Z)]} \qquad (26)$$

$$\varepsilon = \frac{\text{NTU}}{1 + \text{NTU}} \text{ for } Z = 1 \qquad (27)$$

Incropera and DeWitt (1996) and Kays and London (1984) show the relations of ε, NTU, and Z for other flow arrangements. These authors and Afgan and Schlunder (1974) present graphical representations for convenience.

TRANSIENT HEAT FLOW

Often, the heat transfer and temperature distribution under unsteady-state (varying

part I Theory

with time) conditions must be known. Examples are (1) cold storage temperature variations on starting or stopping a refrigeration unit; (2) variation of external air temperature and solar irradiation affecting the heat load of a cold storage room or wall temperatures; (3) the time required to freeze a given material under certain conditions in a storage room; (4) quick freezing of objects by direct immersion in brines; and (5) sudden heating or cooling of fluids and solids from one temperature to a different temperature.

The equations describing transient temperature distribution and heat transfer are presented in this section. Numerical methods are the simplest means of solving these equations because numerical data are easy to obtain. However, with some numerical solutions and off-the-shelf software, the physics that drives the energy transport can be lost. Thus, analytical solution techniques are also included in this section.

The fundamental equation for unsteady-state conduction in solids or fluids in which there is no substantial motion is

$$\frac{\partial t}{\partial \tau} = \alpha \left(\frac{\partial^2 t}{\partial x^2} + \frac{\partial^2 t}{\partial y^2} + \frac{\partial^2 t}{\partial z^2} \right) \tag{28}$$

where thermal diffusivity α is the ratio $k/\rho c_p$; k is thermal conductivity; ρ, density; and c_p, specific heat. If α is large (high conductivity, low density and specific heat, or both), heat will diffuse faster.

One of the most elementary transient heat transfer models predicts the rate of temperature change of a body or material being held at constant volume with uniform temperature, such as a well-stirred reservoir of fluid whose temperature is changing because of a net rate of heat gain or loss:

$$q_{net} = (Mc_p) \frac{dt}{d\tau} \tag{29}$$

where M is the mass of the body, c_p is the specific heat at constant pressure, and q_{net} is the net heat transfer rate to the substance (heat transfer into the substance is positive, and heat transfer out of the substance is negative). Equation (29) is applicable when the pressure around the substance is constant; if the volume of the substance is constant, c_p should be replaced by the constant volume specific heat c_v. It should be noted that with the density of solids and liquids being almost constant, the two specific heats are almost equal. The term q_{net} may include heat transfer by conduction, convection, or radiation and is the difference between the heat transfer rates into and out of the body.

From Equations (28) and (29), it is possible to derive expressions for temperature and heat flow variations at different instants and different locations. Most common cases have been solved and presented in graphical forms (Jakob 1957, Schneider 1964, Myers 1971). In other cases, it is simpler to use numerical methods (Croft and Lilley 1977, Patankar 1980). When convective boundary conditions are required in the solution of Equations (28) and (29), h values based on steady-state correlations are often used. However, this approach may not be valid when rapid transients are involved.

Estimating Cooling Times

Cooling times for materials can be estimated (McAdams 1954) by Gurnie-Lurie charts (Figures 2, 3, and 4), which are graphical solutions for the heating or cooling of infinite slabs, infinite cylinders, and spheres. These charts assume an initial uniform temperature distribution and no change of phase. They apply to a body exposed to a constant temperature fluid with a constant surface convection coefficient of h.

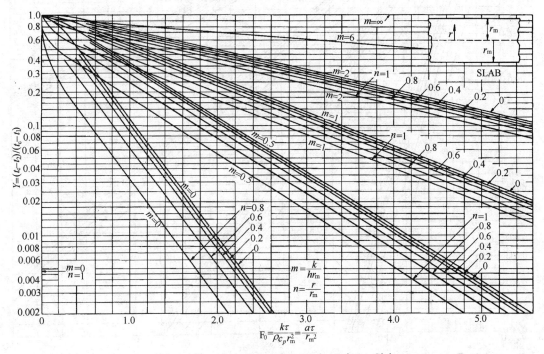

Fig. 2 Transient Temperatures for Infinite Slab

Using Figures 2, 3, and 4, it is possible to estimate both the temperature at any point and the average temperature in a homogeneous mass of material as a function of time in a cooling process. It is possible to estimate cooling times for rectangular-shaped solids, cubes, cylinders, and spheres.

From the point of view of heat transfer, a cylinder insulated on its ends behaves like a cylinder of infinite length, and a rectangular solid insulated so that only two parallel faces allow heat transfer behaves like an infinite slab. A thin slab or a long, thin cylinder may be also considered infinite objects.

Consider a slab having insulated edges being cooled. If the cooling time is the time required for the center of the slab to reach a temperature of t_2, the cooling time can be calculated as follows:

1. Evaluate the temperature ratio $(t_c - t_2)/(t_c - t_1)$.

where

   t_c = temperature of cooling medium

part I Theory

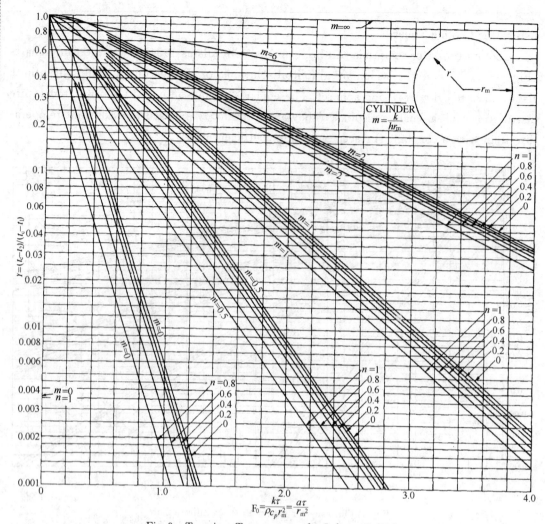

Fig. 3 Transient Temperatures for Infinite Cylinder

t_1 = initial temperature of product

t_2 = final temperature of product at center

Note that in Figures 2, 3, and 4, the temperature ratio $(t_c - t_2)/(t_c - t_1)$ is designated as Y to simplify the equations.

2. Determine the radius ratio r/r_m designated as n in Figures 2, 3, and 4.

where

r = distance from centerline

r_m = half thickness of slab

3. Evaluate the resistance ratio k/hr_m designated as m in Figures 2, 3, and 4.

where

k = thermal conductivity of material

h = heat transfer coefficient

4. From Figure 2 for infinite slabs, select the appropriate value of $k\tau/\rho c_p r_m^2$ designated as

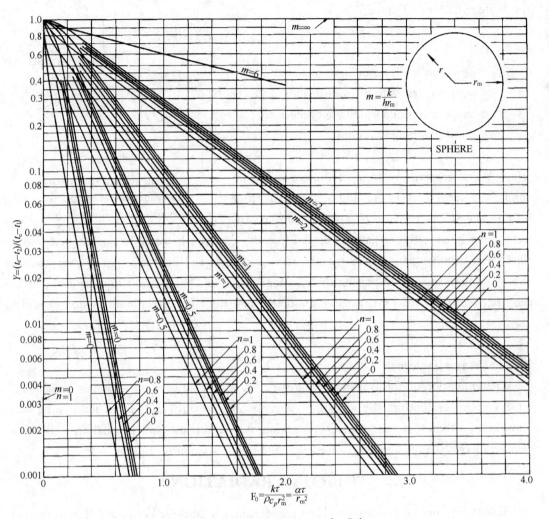

Fig. 4   Transient Temperatures for Spheres

F_0 in Figures 2, 3, and 4.

where

τ = time elapsed

c_p = specific heat

ρ = density

5. Determine τ from the value of $k\tau/\rho c_p r_m^2$.

Multidimensional Temperature Distribution

The solution for semi-infinite slabs and cylinders (shown in Figures 2, 3, and 4) can be used to find the temperatures in finite rectangular solids or cylinders.

The temperature in the finite object can be calculated from the temperature ratio Y of the infinite objects that intersect to form the finite object. The product of the temperature ratios of the infinite objects is the temperature ratio of the finite object; for example, for the finite cylinder of Figure 5,

$$Y_{fc} = Y_{is} Y_{ic} \quad (30)$$

where

Y_{fc} = temperature ratio of finite cylinder
Y_{is} = temperature ratio of infinite slab
Y_{ic} = temperature ratio of infinite cylinder

For a finite rectangular solid,

$$Y_{frs} = (Y_{is})_1 (Y_{is})_2 (Y_{is})_3 \quad (31)$$

where Y_{frs} = temperature ratio of finite rectangular solid, and subscripts 1, 2, and 3 designate three infinite slabs. The convective heat transfer coefficients associated with one pair of parallel surfaces need not be equal to the coefficient associated with another pair. However, the temperature of the fluid adjacent to every surface should be the same. In evaluating the resistance ratio and the Fourier number Fo, the appropriate values of the heat transfer coefficient and the characteristic dimension should be used.

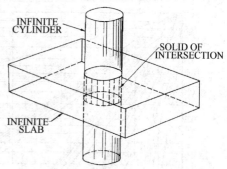

Fig. 5 Finite Cylinder of Intersection from Intersection of Infinite cylinder and Infinite Slab

Heat Exchanger Transients

Determination of the transient behavior of heat exchangers is becoming increasingly important in evaluating the dynamic behavior of heating and air-conditioning systems. Many studies of the transient behavior of counterflow and parallel flow heat exchangers have been conducted; some are listed in the section on Bibliography.

THERMAL RADIATION

Radiation, one of the basic mechanisms for energy transfer between different temperature regions, is distinguished from conduction and convection in that it does not depend on an intermediate material as a carrier of energy but rather is impeded by the presence of material between the regions. The radiation energy transfer process is the consequence of energy-carrying electromagnetic waves that are emitted by atoms and molecules due to changes in their energy content. The amount and characteristics of radiant energy emitted by a quantity of material depend on the nature of the material, its microscopic arrangement, and its absolute temperature. Although rate of energy emission is independent of the surroundings, the **net** energy transfer rate depends on the temperatures and spatial relationships of the surface and its surroundings.

Blackbody Radiation

The rate of thermal radiant energy emitted by a surface depends on its absolute temperature. A surface is called **black** if it can absorb all incident radiation. The total energy emitted per unit time per unit area of black surface W_b to the hemispherical region above it is given by the **Stefan-Boltzmann law**:

$$W_b = \sigma T^4 \tag{32}$$

where W_b is the total rate of energy emission per unit area, and σ is the Stefan-Boltzmann constant $[5.670 \times 10^{-8} \text{W}/(\text{m}^2 \cdot \text{K}^4)]$.

The heat radiated by a body comprises electromagnetic waves of many different frequencies or wavelengths. Planck showed that the spectral distribution of the energy radiated by a blackbody is

$$W_{b\lambda} = \frac{C_1 \lambda^{-5}}{e^{C_2/\lambda T} - 1} \tag{33}$$

where

$W_{b\lambda}$ = monochromatic emissive power of blackbody, W/m^3
λ = wavelength, μm
T = temperature, K
C_1 = first Planck's law constant = $3.742 \times 10^{-16} \text{W} \cdot \text{m}^2$
C_2 = second Planck's law constant = $0.014388 \text{m} \cdot \text{K}$

$W_{b\lambda}$ is the **monochromatic emissive power**, defined as the energy emitted per unit time per unit surface area at wavelength λ per unit wavelength interval around λ; that is, the energy emitted per unit time per unit surface area in the interval $d\lambda$ is equal to $W_{b\lambda} d\lambda$.

The Stefan-Boltzmann equation can be obtained by integrating Planck's equation:

$$W_b = \sigma T^4 = \int_0^\infty W_{b\lambda} d\lambda \tag{34}$$

Wien showed that the wavelength of maximum emissive power multiplied by the absolute temperature is a constant:

$$\lambda_{max} T = 2898 \mu\text{m} \cdot \text{K} \tag{35}$$

where λ_{max} is the wavelength at which the monochromatic emissive power is a maximum and not the maximum wavelength. Equation (35) is known as **Wien's displacement law.** According to this law, the maximum spectral emissive power is displaced to shorter wavelengths with increasing temperature, such that significant emission eventually occurs over the entire visible spectrum as shorter wavelengths become more prominent. For additional details, see Incropera and DeWitt (1996).

Actual Radiation

Substances and surfaces diverge variously from the Stefan-Boltzmann and Planck laws. W_b and $W_{b\lambda}$ are the maximum emissive powers at a surface temperature. Actual surfaces emit and absorb less than these maximums and are called **nonblack.** The emissive power of a nonblack surface at temperature T radiating to the hemispherical region above it is written as

$$W = \varepsilon W_b = \varepsilon \sigma T^4 \tag{36}$$

where ε is known as the **hemispherical emittance.** The term emittance onforms to physical and electrical terminology; the suffix "ance" denotes a property of a piece of material as it exists. The ending "ivity" denotes a property of the bulk material independent of geometry or surface condition. Thus, emittance, reflectance, absorptance, and transmittance refer to actual pieces of material. Emissivity, reflectivity, absorptivity, and transmissivity refer to the properties of materials that are optically smooth and thick enough to be opaque.

The emittance is a function of the material, the condition of its surface, and the temperature of the surface. Table 3 lists selected values; Siegel and Howell (1981) and Modest (1993) have more extensive lists.

Table 3 Emittances and Absorptances for Some Surfaces

Class	Surfaces	Total Normal Emittance[a]		Absorptance for Solar Radiation
		At 10 to 40°C	At 500°C	
1	A small hole in a large box, sphere, furnace, or enclosure	0.97 to 0.99	0.97 to 0.99	0.97 to 0.99
2	Black nonmetallic surfaces such as asphalt, carbon, slate, paint, paper	0.90 to 0.98	0.90 to 0.98	0.85 to 0.98
3	Red brick and tile, concrete and stone, rusty steel and iron, dark paints (red, brown, green, etc.)	0.85 to 0.95	0.75 to 0.90	0.65 to 0.80
4	Yellow and buff brick and stone, firebrick, fireclay	0.85 to 0.95	0.70 to 0.85	0.50 to 0.70
5	White or light cream brick, tile, paint or paper, plaster, whitewash	0.85 to 0.95	0.60 to 0.75	0.30 to 0.50
6	Window glass	0.90	—	b
7	Bright aluminum paint; gilt or bronze paint	0.40 to 0.60	—	0.30 to 0.50
8	Dull brass, copper, or aluminum; galvanized steel; polished iron	0.20 to 0.30	0.30 to 0.50	0.40 to 0.65
9	Polished brass, copper, monel metal	0.02 to 0.05	0.05 to 0.15	0.30 to 0.50
10	Highly polished aluminum, tin plate, nickel, chromium	0.02 to 0.04	0.05 to 0.10	0.10 to 0.40
11	Selective surfaces			
	Stainless steel wire mesh	0.23 to 0.28	—	0.63 to 0.86
	White painted surface	0.92	—	0.23 to 0.49
	Copper treated with solution of $NaClO_2$ and $NaOH$	0.13	—	0.87
	Copper, nickel, and aluminum plate with CuO coating	0.09 to 0.21	—	0.08 to 0.93

[a] Hemispherical and normal emittance are unequal in many cases. The hemispherical emittance may vary from up to 30% greater for polished reflectors to 7% lower for nonconductors.

[b] Absorbs 4 to 40% depending on its transmittance.

The monochromatic emissive power of a nonblack surface is similarly written as

$$W_\lambda = \varepsilon_\lambda W_{b\lambda} = \varepsilon_\lambda \left(\frac{C_1 \lambda^{-5}}{e^{C_2/\lambda T} - 1} \right) \tag{37}$$

where ε_λ is the monochromatic hemispherical emittance. The relationship between ε and ε_λ

is given by

$$W = \varepsilon \sigma T^4 = \int_0^\infty W_\lambda d\lambda = \int_0^\infty \varepsilon_\lambda W_{b\lambda} d\lambda$$

or

$$\varepsilon = \frac{1}{\sigma T^4} \int_0^\infty \varepsilon_\lambda W_{b\lambda} d\lambda \tag{38}$$

If ε_λ does not depend on λ, then, from Equation (38), $\varepsilon = \varepsilon_\lambda$. Surfaces with this characteristic are called **gray**. Gray surface characteristics are often assumed in calculations. Several classes of surfaces approximate this condition in some regions of the spectrum. The simplicity is desirable, but care must be exercised, especially if temperatures are high. Assumption of grayness is sometimes made because of the absence of information relating ε_λ and λ.

When radiant energy falls on a surface, it can be absorbed, reflected, or transmitted through the material. Therefore, from the first law of thermodynamics,

$$\alpha + \tau + \rho = 1 \tag{39}$$

where

α = fraction of incident radiation absorbed or **absorptance**

τ = fraction of incident radiation transmitted or **transmittance**

ρ = fraction of incident radiation reflected or **reflectance**

If the material is opaque, as most solids are in the infrared, $\tau = 0$ and $\alpha + \rho = 1$. For a black surface, $\alpha = 1$, $\rho = 0$, and $\tau = 0$. Platinum black and gold black are as black as any actual surface and have absorptances of about 98% in the infrared. Any desired degree of blackness can be simulated by a small hole in a large enclosure. Consider a ray of radiant energy entering the opening. It will undergo many internal reflections and be almost completely absorbed before it has a reasonable probability of passing back out of the opening.

Certain flat black paints also exhibit emittances of 98% over a wide range of conditions. They provide a much more durable surface than gold or platinum black and are frequently used on radiation instruments and as standard reference in emittance or reflectance measurements.

Kirchhoff's law relates emittance and absorptance of any opaque surface from thermodynamic considerations; it states that for any surface where the incident radiation is independent of angle or where the surface is diffuse, $\varepsilon_\lambda = \alpha_\lambda$. If the surface is gray, or the incident radiation is from a black surface at the same temperature, then $\varepsilon = \alpha$ as well, but many surfaces are not gray. For most surfaces listed in Table 3, absorptance for solar radiation is different from emittance for low-temperature radiation. This is because the wavelength distributions are different in the two cases, and ε_λ varies with wavelength.

The foregoing discussion relates to total hemispherical radiation from surfaces. Energy distribution over the hemispherical region above the surface also has an important effect on the rate of heat transfer in various geometric arrangements.

Lambert's law states that the emissive power of radiant energy over a hemispherical surface above the emitting surface varies as the cosine of the angle between the normal to the radiating surface and the line joining the radiating surface to the point of the hemispherical surface. This radiation is **diffuse radiation.** The Lambert emissive power variation is equivalent to assuming that radiation from a surface in a direction other than normal occurs as if it came from an equivalent area with the same emissive power (per unit area) as the original surface. The equivalent area is obtained by projecting the original area onto a plane normal to the direction of radiation. Black surfaces obey the Lambert law exactly. The law is approximate for many actual radiation and reflection processes, especially those involving rough surfaces and nonmetallic materials. Most radiation analyses are based on the assumption of gray diffuse radiation and reflection.

In estimating heat transfer rates between surfaces of different geometries, radiation characteristics, and orientations, it is usually assumed that
- All surfaces are gray or black
- Radiation and reflection are diffuse
- Properties are uniform over the surfaces
- Absorptance equals emittance and is independent of the temperature of the source of incident radiation
- The material in the space between the radiating surfaces neither emits nor absorbs radiation

These assumptions greatly simplify problems, although results must be considered approximate.

Angle Factor

The distribution of radiation from a surface among the surfaces it irradiates is indicated by a quantity variously called an interception, a view, a configuration, a shape factor, or an angle factor. In terms of two surfaces i and j, the **angle factor** F_{ij} from surface i to surface j is the ratio of the radiant energy leaving surface i and directly reaching surface j to the total radiant energy leaving surface i. The angle factor from j to i is similarly defined, merely by interchanging the roles of i and j. This second angle factor is not, in general, numerically equal to the first. However, the reciprocity relation $F_{ij}A_i = F_{ji}A_j$, where A is the surface area, is always valid. Note that a concave surface may "see itself" ($F_{ii} \neq 0$), and that if n surfaces form an enclosure,

$$\sum_{j=1}^{n} F_{ij} = 1 \tag{40}$$

The angle factor F_{12} between two surfaces is

$$F_{12} = \frac{1}{A_1} \int_{A_1} \int_{A_2} \frac{\cos\phi_1 \cos\phi_2}{\pi r^2} dA_1 dA_2 \tag{41}$$

where dA_1, and dA_2 are elemental areas of the two surfaces, r is the distance between dA_1 and dA_2, and ϕ_1 and ϕ_2 are the angles between the respective normals to dA_1 and dA_2 and the connecting line r. Numerical, graphical, and mechanical techniques can solve this equation (Siegel and Howell 1981, Modest 1993). Numerical values of the angle factor for common geometries are given in Figure 6.

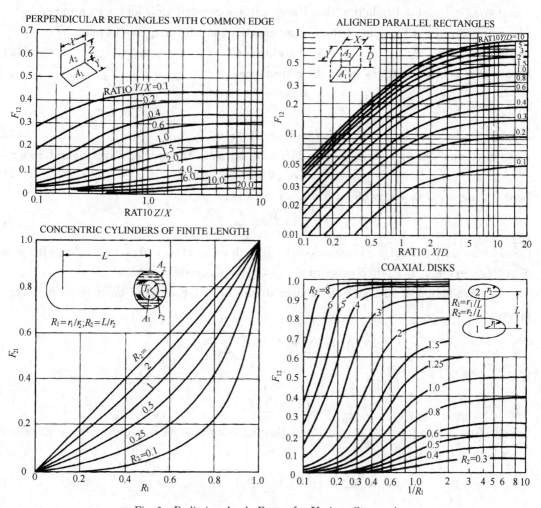

Fig. 6 Radiation Angle Factor for Various Geometries

Calculation of Radiant Exchange Between Surfaces Separated by Nonabsorbing Media

A surface radiates energy at a rate independent of its surroundings and absorbs and reflects incident energy at a rate dependent on its surface condition. The net energy exchange per unit area is denoted by q or q_j for unit area A_j. It is the rate of emission of the surface minus the total rate of absorption at the surface from all radiant effects in its surroundings, possibly including the return of some of its own emission by reflection off its surroundings. The rate at which energy must be supplied to the surface by other exchange processes if its temperature is to remain constant is q; therefore, to define q, the total radiant surroundings (in effect, an enclosure) must be specified.

Several methods have been developed to solve certain problems. To calculate the radiation exchange at each surface of an enclosure of n opaque surfaces by simple, general equations convenient for machine calculation, two terms must be defined:

$G =$ irradiation; total radiation incident on surface per unit time and per unit area

$J =$ radiosity; total radiation that leaves surface per unit time and per unit area

The radiosity is the sum of the energy emitted and the energy reflected:

$$J = \varepsilon W_b + \rho G \tag{42}$$

Because the transmittance is zero, the reflectance is

$$\rho = 1 - \alpha = 1 - \varepsilon$$

Thus,

$$J = \varepsilon W_b + (1-\varepsilon) G \tag{43}$$

The net energy lost by a surface is the difference between the radiosity and the irradiation:

$$q/A = J - G = \varepsilon W_b + (1-\varepsilon) G - G \tag{44}$$

Substituting for G in terms of J from Equation (43),

$$q = \frac{W_b - J}{(1-\varepsilon)/\varepsilon A} \tag{45}$$

Consider an enclosure of n isothermal surfaces with areas of A_1, A_2, $\cdots$, A_n, emittances of ε_1, ε_2, $\cdots$, ε_n, and reflectances of ρ_1, ρ_2, $\cdots$, ρ_n, respectively.

The irradiation of surface i is the sum of the radiation incident on it from all n surfaces:

$$G_i A_i = \sum_{j=1}^{n} F_{ji} J_j A_j = \sum_{j=1}^{n} F_{ij} J_j A_i$$

or

$$G_i = \sum_{j=1}^{n} F_{ij} J_j$$

Substituting in Equation (44) yields the following simultaneous equations when each of the n surfaces is considered:

$$J_i = \varepsilon_i W_{bi} + (1-\varepsilon_i) \sum_{j=1}^{n} F_{ij} J_j \quad i = 1, 2, \cdots, n \tag{46}$$

Equation (46) can be solved manually for the unknown J_s if the number of surfaces is small. The solution for more complex enclosures requires a computer.

Once the radiosities (J_s) are known, the net radiant energy lost by each surface is determined from Equation (45) as

$$q_i = \frac{W_{bj} - J_i}{(1-\varepsilon_i)/\varepsilon_i A_i}$$

If the surface is black, Equation (45) becomes indeterminate, and an alternate expression must be used, such as

$$q_i = \sum_{j=1}^{n} J_i A_i F_{ij} - J_j A_j F_{ji}$$

or

$$q_i = \sum_{j=1}^{n} F_{ij} A_i (J_i - J_j) \qquad (47)$$

since

$$F_{ij} A_i = F_{ji} A_j$$

All diffuse radiation processes are included in the aforementioned enclosure method, and surfaces with special characteristics are assigned consistent properties. An opening is treated as an equivalent surface area A_e with a reflectance of zero. If energy enters the enclosure diffusely through the opening, A_e is assigned an equivalent temperature; otherwise, its temperature is taken as zero. If the loss through the opening is desired, q_2 is found. A window in the enclosure is assigned its actual properties.

A surface in **radiant balance** is one for which radiant emission is balanced by radiant absorption; heat is neither removed from nor supplied to the surface. Reradiating surfaces (insulated surfaces with $q_{net} = 0$), can be treated in Equation (46) as being perfectly reflective (i.e., $\varepsilon = 0$). The equilibrium temperature of such a surface can be found from

$$T_k = \left(\frac{J_k}{\sigma}\right)^{0.25}$$

once Equation (46) has been solved for the radiosities.

Use of angle factors and radiation properties as defined assumes that the surfaces are diffuse radiators—a good assumption for most nonmetals in the infrared region, but a poor assumption for highly polished metals. Subdividing the surfaces and considering the variation of radiation properties with angle of incidence improves the approximation but increases the work required for a solution.

Radiation in Gases

Elementary gases such as oxygen, nitrogen, hydrogen, and helium are essentially transparent to thermal radiation. Their absorption and emission bands are confined mainly to the ultraviolet region of the spectrum. The gaseous vapors of most compounds, however, have absorption bands in the infrared region. Carbon monoxide, carbon dioxide, water vapor, sulfur dioxide, ammonia, acid vapors, and organic vapors absorb and emit significant amounts of energy.

Radiation exchange by opaque solids is considered a surface phenomenon. Radiant energy does, however, penetrate the surface of all materials. The absorption coefficient gives the rate of exponential attenuation of the energy. Metals have large absorption coefficients, and radiant energy penetrates less than 100 nm at most. Absorption coefficients for nonmetals are lower. Radiation may be considered a surface phenomenon unless the material is transparent. Gases have small absorption coefficients, so the path length of radiation through gas becomes very significant.

Beer's law states that the attenuation of radiant energy in a gas is a function of the product $p_g L$ of the partial pressure of the gas and the path length. The monochromatic absorptance of a body of gas of thickness L is then given by

$$\alpha_{\lambda L} = 1 - e^{-\alpha \lambda L} \tag{48}$$

Because absorption occurs in discrete wavelengths, the absorptances must be summed over the spectral region corresponding to the temperature of the blackbody radiation passing through the gas. The monochromatic absorption coefficient α_λ is also a function of temperature and pressure of the gas; therefore, detailed treatment of gas radiation is quite complex.

Estimated emittance for carbon dioxide and water vapor in air at 24 ℃ is a function of concentration and path length (Table 4). The values are for a hemispherically shaped body of gas radiating to an element of area at the center of the hemisphere. Among others, Modest (1993), Siegel and Howell (1981), and Hottel and Sarofim (1967) describe geometrical calculations in their texts on radiation transfer. Generally, at low values of $p_g L$, the mean path length L (or equivalent hemispherical radius for a gas body radiating to its surrounding surfaces) is four times the mean hydraulic radius of the enclosure. A room with a dimensional ratio of 1 : 1 : 4 has a mean path length of 0.89 times the shortest dimension when considering radiation to all walls. For a room with a dimensional ratio of 1 : 2 : 6, the mean path length for the gas radiating to all surfaces is 1.2 times the shortest dimension. The mean path length for radiation to the 2 by 6 face is 1.18 times the shortest dimension. These values are for cases where the partial pressure of the gas times the mean path length approaches zero ($p_g L \approx 0$). The factor decreases with increasing values of $p_g L$. For average rooms with approximately 2.4 m ceilings and relative humidity ranging from 10 to 75% at 24 ℃, the effective path length for carbon dioxide radiation is about 85% of the ceiling height, or 2.0 m. The effective path length for water vapor is about 93% of the ceiling height, or 2.3 m. The effective emittance of the water vapor and carbon dioxide radiating to the walls, ceiling, and floor of a room 4.9 m by 14.6 m with 2.4 m ceilings is in the following tabulation.

Emittance of CO₂ and Water Vapor in Air at 24 ℃ Table 4

Path Length, m	CO_2, % by Volume			Relative Humidity, %		
	0.1	0.3	1.0	10	50	100
	0.03	0.06	0.09	0.06	0.17	0.22
30	0.09	0.12	0.16	0.22	0.39	0.47
300	0.16	0.19	0.23	0.47	0.64	0.70

Relative Humidity, %	ε_g	Relative Humidity, %	ε_g
10	0.10	75	0.22
50	0.19		

The radiation heat transfer from the gas to the walls is then

$$q = \sigma A_w \varepsilon_g (T_g^4 - T_w^4) \tag{49}$$

The examples in Table 4 and the preceding text indicate the importance of gas radiation in environmental heat transfer problems. Gas radiation in large furnaces is the dominant mode of heat transfer, and many additional factors must be considered. Increased pressure broadens the spectral bands, and interaction of different radiating species prohibits simple summation of the emittance factors for the individual species. Departures from blackbody conditions necessitate separate calculations of the emittance and absorptance. McAdams (1954) and Hottel and Sarofim (1967) give more complete treatments of gas radiation.

NATURAL CONVECTION

Heat transfer involving motion in a fluid due to the difference in density and the action of gravity is called **natural convection** or **free convection**. Heat transfer coefficients associated with gases for natural convection are generally much lower than those for forced convection, and it is therefore important not to ignore radiation in calculating the total heat loss or gain. Radiant transfer may be of the same order of magnitude as natural convection, even at roomtemperatures, because wall temperatures in a room can affect human comfort.

Natural convection is important in a variety of heating and refrigeration equipment: (1) gravity coils used in high-humidity cold storage rooms and in roof-mounted refrigerant condensers, (2) the evaporator and condenser of household refrigerators, (3) baseboard radiators and convectors for space heating, and (4) cooling panels for air conditioning. Natural convection is also involved in heat loss or gain to equipment casings and interconnecting ducts and pipes.

Consider heat transfer by natural convection between a cold fluid and a hot surface. The fluid in immediate contact with the surface is heated by conduction, becomes lighter, and rises because of the difference in density of the adjacent fluid. The viscosity of the fluid resists this motion. The heat transfer is influenced by (1) gravitational force due to thermal expansion, (2) viscous drag, and (3) thermal diffusion. Gravitational acceleration g, coefficient of thermal expansion β, kinematic viscosity $\nu = \mu/\rho$, and thermal diffusivity $\alpha = k/\rho c_p$ affect natural convection. These variables are included in the dimensionless numbers given in Equation (1) in Table 5. The Nusselt number Nu is a function of the product of the Prandtl number Pr and the Grashof number Gr. These numbers, when combined, depend on the fluid properties, the temperature difference Δt between the surface and the fluid, and the characteristic length L of the surface. The constant c and the exponent n depend on the physical configuration and the nature of flow.

Natural convection cannot be represented by a single value of exponent n, but it can be divided into three regions:

part I Theory

Natural Convection Heat Transfer Coefficients Table 5

I. General relationships

$$Nu = c(Gr\,Pr)^n \quad (1)$$

$$h = c\frac{k}{L}\left(\frac{L^3\rho^2\beta g\Delta t}{\mu^2}\right)_f^n \left(\frac{\mu c_p}{k}\right)_f^n \quad (2)$$

Characteristic length L

Vertical plates or pipes	L = height
Horizontal plates	L = length
Horizontal pipes	L = diameter
Spheres	$L = 0.5 \times$ diameter
Rectangular block, with horizontal length L_h and vertical length L_v	$1/L = (1/L_h) + (1/L_v)$

II. Planes and pipes

Horizontal or vertical planes, pipes, rectangular blocks, and spheres (excluding horizontal plates facing downward for heating and upward for cooling)

(a) Laminar range, when Gr Pr is between 10^4 and 10^8 $\quad Nu = 0.56(Gr\,Pr)^{0.25} \quad (3)$

(b) Turbulent range, when Gr Pr is between 10^8 and 10^{12} $\quad Nu = 0.13(Gr\,Pr)^{0.33} \quad (4)$

III. Wires

For horizontal or vertical wires, use L = diameter, for Gr Pr between 10^{-7} and 1 $\quad Nu = (Gr\,Pr)^{0.1} \quad (5)$

IV. With air

Gr Pr $= 1.6 \times 10^6 L^3 \Delta t$ (at 21℃, L in m, Δt in K)

(a) Horizontal cylinders

    Small cylinder, laminar range $\quad h = 1.32(\Delta t/L)^{0.25} \quad (6)$

    Large cylinder, turbulent range $\quad h = 1.24(\Delta t)^{0.33} \quad (7)$

(b) Vertical plates

    Small plates, laminar range $\quad h = 1.42(\Delta t/L)^{0.25} \quad (8)$

    Large plates, turbulent range $\quad h = 1.31(\Delta t)^{0.33} \quad (9)$

(c) Horizontal plates, facing upward when heated or downward when cooled Small plates, laminar range $\quad h = 1.32(\Delta t/L)^{0.25} \quad (10)$

    Large plates, turbulent range $\quad h = 1.52(\Delta t)^{0.33} \quad (11)$

(d) Horizontal plates, facing downward when heated or upward when cooled Small plates $\quad h = 0.59(\Delta t/L)^{0.25} \quad (12)$

1. **Turbulent** natural convection, for which n equals 0.33
2. **Laminar** natural convection, for which n equals 0.25
3. A region that has GrPr less than for laminar natural convection, for which the exponent n gradually diminishes from 0.25 to lower values

Note that for wires, the GrPr is likely to be very small, so that the exponent n is 0.1 [Equation (5) in Table 5].

    To calculate the natural-convection heat transfer coefficient, determine GrPr to find whether the boundary layer is laminar or turbulent; then apply the appropriate equation from Table 5. The correct characteristic length indicated in the table must be used. Because the exponent n is 0.33 for a turbulent boundary layer, the characteristic length cancels out in Equation (2) in Table 5, and the heat transfer coefficient is independent of the characteristic length, as seen in Equations (7), (9), and (11) in Table 5. Turbulence occurs when length or temperature difference is large. Because the length of a pipe is generally greater than its diameter, the heat transfer coefficient for vertical pipes is larger than for horizontal pipes.

Convection from horizontal plates facing downward when heated (or upward when cooled) is a special case. Because the hot air is above the colder air, theoretically no convection should occur. Some convection is caused, however, by secondary influences such as temperature differences on the edges of the plate. As an approximation, a coefficient of somewhat less than half the coefficient for a heated horizontal plate facing upward can be used.

Because air is often the heat transport fluid, simplified equations for air are given in Table 5. Other information on natural convection is available in the section on Bibliography under Heat Transfer, General.

<u>Observed differences in the comparison of recent experimental and numerical results with existing correlations for natural convective heat transfer coefficients indicate that caution should be used when applying coefficients for (isolated) vertical plates to vertical surfaces in enclosed spaces (buildings).</u> Bauman et al. (1983) and Altmayer et al. (1983) developed improved correlations for calculating natural convective heat transfer from vertical surfaces in rooms under certain temperature boundary conditions.

Natural convection can affect the heat transfer coefficient in the presence of weak forced convection. As the forced-convection effect (i. e., the Reynolds number) increases, "mixed convection" (super-imposed forced-on-free convection) gives way to the pure forced-convection regime. In these cases, other sources describing combined free and forced convection should be consulted, since the heat transfer coefficient in the mixed-convection region is often larger than that calculated based on the natural- or forced-convection calculation alone. Metais and Eckert (1964) summarize natural-, mixed-, and forced-convection regimes for vertical and horizontal tubes. Figure 7 shows the approximate limits for horizontal

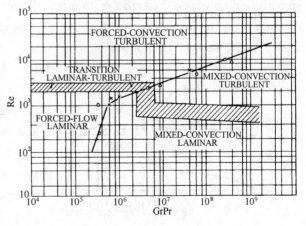

Fig. 7 Regimes of Free, Forced, and Mixed Convection for flow-Through Horizontal Tubes

tubes. Other studies are described by Grigull et al. (1982).

FORCED CONVECTION

Forced air coolers and heaters, forced air- or water-cooled condensers and evaporators, and liquid suction heat exchangers are examples of equipment that transfer heat primarily by forced convection.

When fluid flows over a flat plate, a **boundary layer** forms adjacent to the plate. The velocity of the fluid at the plate surface is zero and increases to its maximum free stream value just past the edge of the boundary layer (Figure 8). Boundary layer formation is important because the temperature change from plate to fluid (thermal resistance) is concentrated here. Where the boundary layer is thick, thermal resistance is great and the heat transfer coefficient is small. Flow within the boundary layer immediately downstream from the leading edge is laminar and is known as **laminar forced convection**. As flow proceeds along the plate, the laminar boundary layer increases in thickness to a critical value. Then, turbulent eddies develop within the boundary layer, except for a thin **laminar sublayer** adjacent to the plate.

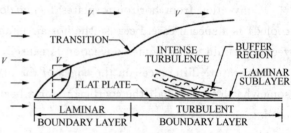

Fig. 8 Boundary Layer Buildup on Flat Plate
(Vertical Scale Magnified)

The boundary layer beyond this point is a **turbulent boundary layer**, and the flow is **turbulent forced convection**. The region between the breakdown of the laminar boundary layer and the establishment of the turbulent boundary layer is the **transition region**. Because the turbulent eddies greatly enhance heat transport into the main stream, the heat transfer coefficient begins to increase rapidly through the transition region. For a flat plate with a smooth leading edge, the turbulent boundary layer starts at Reynolds numbers, based on distance from the leading edge, of about 300000 to 500000. In blunt-edged plates, it can start at much smaller Reynolds numbers.

For long tubes or channels of small hydraulic diameter, at sufficiently low flow velocity, the laminar boundary layers on each wall grow until they meet. Beyond this point, the velocity distribution does not change, and no transition to turbulent flow takes place. This is called **fully developed laminar flow**. For tubes of large diameter or at higher velocities, transition to turbulence takes place and **fully developed turbulent flow** is established (Figure 9). Therefore, the length dimension that determines the critical Reyn-

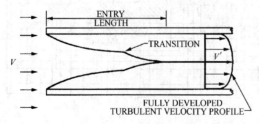

Fig. 9 Boundary Layer Buildup in Entry Length of Tube or Channel

olds number is the hydraulic diameter of the channel. For smooth circular tubes, flow is laminar for Reynolds numbers below 2100 and turbulent above 10000.

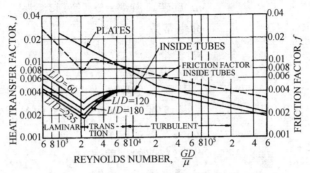

Fig. 10　Typical Dimensionless Representation of Forced-Convection Heat Transfer

WORDS AND EXPRESSIONS

address	*vt.* 向…致辞，演说，写姓名地址，从事，忙于；*n.* 地址，致辞，演讲，说话的技巧
aforementioned	*adj.* 上述的，前述的
be analogous to	与……相类比
be designated as…	被定义为……
be directly proportional to	和……成正比
be inversely proportional to	和……成反比
be termed……	被称为……
blower	*n.* 吹制工，送风机，吹风机，＜俚＞爱吹牛的人
brass	*n.* 黄铜，黄铜制品，厚脸皮
brine	*n.* 盐水
bronze	*n.* 青铜(铜与锡合金)，铜像；*adj.* 青铜色的
buffer layer	过渡流层
characteristic length	特征长度
chilled water system	冷冻水系统
circular	*adj.* 圆形的，循环的；*n.* 函件，通知
coil	*v.* 盘绕，卷；*n.* 盘管
concave	*adj.* 凹的
condenser	*n.* 冷凝器，电容器
consistent	*adj.* 一致的，调和的，坚固的，[数、统] 相容的
convex	凸的
depict	*vt.* 描述，描写
diffuse radiation	漫辐射

part I *Theory*

diffusivity	*n.* 扩散能力，扩散率
diverge	*vi.* (道路等)分叉，(意见等)分歧，脱离
eddy	*n.* 旋转，漩涡；*v.* (使)起漩涡
electrical permittivity	介电常数
electrode	*n.* 电极
electromagnetic	*adj.* 电磁的
emissivity	*n.* [物] 发射率
emittance	*n.* [原物] 发射度，[热] 辐射本领
evaporator	*n.* 蒸发器，脱水器
foregoing	*adj.* 在前的，前述的
galvanize	*v.* 通电流于，电镀
gilt	*n.* 镀金，表面装饰，小母猪；*adj.* 镀金的
gravity coils	自然冷却盘管
hemispherical emittance	半球发射率
hexagonal	*adj.* 六角形的，六边形的
incident radiation	入射辐射
indenting	*adj.* 成穴的
infinite slab	无限大平板
leading edge of the plate	平板起始端
momentum	*n.* 动量
monochromatic	*adj.* [物] 单色的，单频的
nonhomogeneous	*adj.* 非齐(次，性)的，不[非] 均匀的，非均态的，多相的
normal	*n.* 法线
opaque	*n.* 不透明物；*adj.* 不透明的，不传热的，迟钝的
oscillation	*n.* 摆动，振动
parallel flow	顺流（counter flow 逆流）
prominent	*adj.* 卓越的，显著的，突出的
proportionality factor	比例系数
radiosity	*n.* 有效辐射
reciprocal	*adj.* 互惠的，相应的，倒数的，彼此相反的；*n.* 倒数，互相起作用的事物
reflectance	*n.* 反射比，反射系数
reflectivity	*n.* 反射率
retrofit device	改型装置
spectrum	*n.* 光，光谱，型谱，频谱
specular reflection	镜面反射
suction	*n.* 抽气(水)，吸气(水)
suffix	*n.* [语] 后缀，下标；*vt.* 添后缀 (prefix 前缀)

superimposed		*adj.* 成阶层的，有层理的
superscript		上标（subscript 下标）
terminology		*n.* 术语学
total normal emittance		全波长法向发射率
trail-and-error calculation		试算
transition		*n.* 过渡区
transmittance		*n.* 透射系数
trial-and-error		*n.* 反复实验
trial-and-error calculation		试凑计算
twisted strip		*n.* 扭曲片

NOTATIONS

(1) 表示"在某种情况下"的短语：
　　for this case, in such an instance, in model 1, in case 1, for this scenario
　　表示"表明，说明"的单词和短语：
　　state, manifest, show, indicate, address, illustrate, demonstrate
　　表示"由此看出"的用法：
　　From ......, it is seen that, it concludes that, it yields that, it derives that
　　表示"组成"的单词和短语：
　　comprise of, consist of, be composed of, be made of, constitute

(2) 注意以下用法并熟练掌握：

This chapter <u>presents</u>(介绍) the elementary principles of single-phase heat transfer <u>with emphasis on</u>(重点在于) heating, refrigerating, and air conditioning.

Thermal energy transfer occurs <u>in the direction of</u>(沿……方向)decreasing temperature, <u>a consequence of</u>(由于……的结果)the second law of thermodynamics.

When <u>fluid currents</u>(流体流)are produced by <u>external sources</u>(外源)(for example, a blower or pump), the solid-to-fluid heat transfer <u>is termed</u>(被称为)forced convection.

Mathematical solutions to many heat conduction problems <u>are addressed by</u>(由……提出) Carslaw and Jaeger (1959).

It is defined <u>as follows</u>(如下)：

$$q = UA(t_1 - t_2)$$

<u>where</u>(其中)A is the surface area, t_1 and t_2 are the temperatures of two fluids, <u>respectively</u>(分别).

The contact resistance between the insulation and the pipe wall <u>is assumed negligible</u>(假设可忽略).

As shown above（如上所述）
For a given length L of pipe（对于给定的管长 L）
be obtained from（从……获得）

be established by（通过……建立）

It is usually assumed（经常假设）

(3) Pronunciation of Formula（数学公式的读法）

$$\varepsilon = \frac{1}{\sigma T^4} \int_0^\infty \varepsilon_\lambda W_{b\lambda} d\lambda$$

can be read "ε equals the reciprocal of σ times T to the 4 power, multiplied by integral from zero to (positive) infinite of ε_λ times $W_{b\lambda}$ times $d\lambda$".

$$W_\lambda = \varepsilon_\lambda \left(\frac{C_1 \lambda^{-5}}{e^{C_2/\lambda T} - 1} \right)$$

can be read "W_λ equals C_1 times λ to the minus 5 power, divided by e to the C_2 over λ and T power minus 1, all multiplied by ε_λ".

(4) One of the most elementary transient heat transfer models predicts the rate of temperature change of a body or material being held at constant volume with uniform temperature, such as a well-stirred reservoir of fluid whose temperature is changing because of a net rate of heat gain or loss.

一个最基本的瞬态传热模型能预测保持在均匀温度的恒定体积的物体或材料的温度变化率，如由于净得热率或净失热率引起充分搅拌的蓄池中流体的温度是变化着的。

(5) Use of angle factors and radiation properties as defined assumes that the surfaces are diffuse radiators—a good assumption for most nonmetals in the infrared region, but a poor assumption for highly polished metals.

角系数和辐射特性如所定义的那样使用是假设这些表面为漫射辐射体——对大多数非金属在红外区假设适用，但对高度磨光的金属假设不太适用。

(6) Consider heat transfer by natural convection between a cold fluid and a hot surface. The fluid in immediate contact with the surface is heated by conduction, becomes lighter, and rises because of the difference in density of the adjacent fluid.

考虑冷流体和热表面间的自然对流换热。流体和表面直接接触时由热传导而被加热，从而变轻，又因为和周围流体的密度差而上升。

EXERCISES

(1) 请翻译以下句子，弄清以"-ance"和"-vity"结尾的英语单词的不同物理含义。

The term emittance conforms to physical and electrical terminology; the suffix "ance" denotes a property of a piece of material as it exists. The ending "ivity" denotes a property of the bulk material independent of geometry of surface condition. Thus, emittance, reflectance, absorptance, and transmittance refer to actual pieces of material. Emissivity, reflectivity, absorptivity, and transmissivity refer to the properties of materials that are optically smooth and thick enough to be opaque.

(2) 请翻译下列句子。

1) Radiation is not significant on the water side because liquids are sensibly opaque to radi-

ation, although water transmits energy in the visible region.

2) The lambert emissive power variation is equivalent to assuming that radiation from a surface in a direction other than normal occurs as if it came from an equivalent area with the same emissive power (per unit area) as the original surface.

3) Observed differences in the comparison of recent experimental and numerical results with existing correlations for natural convection heat transfer coefficients indicate that caution should be used when applying coefficients for (isolated) vertical plates to vertical surfaces in enclosed spaces (building).

(3) 请用英语表述。

并联，串联，

顺流，逆流，

热阻，对数平均温差，

稳态，瞬态，

发射，吸收，

环形翅片，直形翅片，针形翅片

LESSON 4

MASS TRANSFER

MASS transfer by either molecular diffusion or convection is the transport of one component of a mixture relative to the motion of the mixture and is the result of a **concentration gradient.** In an air-conditioning process, water vapor is added or removed from the air by a simultaneous transfer of heat and mass (water vapor) between the airstream and a wetted surface. The wetted surface can be water droplets in an air washer, wetted slats of a cooling tower, condensate on the surface of a dehumidifying coil, surface presented by a spray of liquid absorbent, or wetted surfaces of an evaporative condenser. The performance of equipment with these phenomena must be calculated carefully because of the simultaneous heat and mass transfer.

This chapter is divided into (1) the principles of molecular diffusion and (2) a discussion on the convection of mass.

MOLECULAR DIFFUSION

Most mass transfer problems can be analyzed by considering the diffusion of a gas into a second gas, a liquid, or a solid. In this chapter, the diffusing or dilute component is designated as component B, and the other component as component A. For example, when water vapor diffuses into air, the water vapor is component B and dry air is component A. Properties with subscripts A or B are local properties of that component. Properties without subscripts are local properties of the mixture.

The primary mechanism of mass diffusion at ordinary temperature and pressure conditions is **molecular diffusion**, a result of density gradient. In a binary gas mixture, the presence of a concentration gradient causes transport of matter by molecular diffusion; that is, because of random molecular motion, gas B diffuses through the mixture of gases A and B in a direction that reduces the concentration gradient.

Fick's Law

The basic equation for molecular diffusion is Fick's law. Expressing the concentration of component B of a binary mixture of components A and B in terms of the mass fraction ρ_B/ρ or mole fraction C_B/C, Fick's law is

---

Extracted from Chapter 5 of the ASHRAE *Handbook-Fundamentals.*

$$J_B = -\rho D_v \frac{d(\rho_B/\rho)}{di} = -J_A \tag{1a}$$

$$J_B^* = -CD_v \frac{d(C_B/C)}{di} = -J_A^* \tag{1b}$$

where $\rho = \rho_A + \rho_B$ and $C = C_A + C_B$.

The minus sign indicates that the concentration gradient is negative in the direction of diffusion. The proportionality factor D_v is the **mass diffusivity** or the **diffusion coefficient**. The total mass flux $\dot{m}_B''$ and molar flux $\dot{m}_B''^*$ are due to the average velocity of the mixture plus the diffusive flux:

$$\dot{m}_B'' = \rho_B v - \rho D_v \frac{d(\rho_B/\rho)}{dy} \tag{2a}$$

$$\dot{m}_B''^* = C_B v^* - CD_v \frac{d(C_B/C)}{dy} \tag{2b}$$

where v is the mass average velocity of the mixture and v^* is the molar average velocity.

Bird et al. (1960) present an analysis of Equations (1a) and (1b). Equations (1a) and (1b) are equivalent forms of Fick's law. The equation used depends on the problem and individual preference. This chapter emphasizes mass analysis rather than molar analysis. However, all results can be converted to the molar form using the relation $C_B \equiv \rho_B/M_B$.

Fick's Law for Dilute Mixtures

In many mass diffusion problems, component B is dilute; the density of component B is small compared to the density of the mixture. In this case, Equation (1a) can be written as

$$J_B = -D_v \frac{d\rho_B}{dy} \tag{3}$$

when $\rho_B \ll \rho$ and $\rho_A \approx \rho$.

Equation (3) can be used without significant error for water vapor diffusing through air at atmospheric pressure and a temperature less than 27℃. In this case, $\rho_B < 0.02\rho$, where ρ_B is the density of water vapor and ρ is the density of moist air (air and water vapor mixture). The error in J_B caused by replacing $\rho [d(\rho_B/\rho)/dy]$ with $d\rho_B/dy$ is less than 2%. At temperatures below 60℃ where $\rho_B < 0.10\rho$, Equation (3) can still be used if errors in J_B as great as 10% are tolerable.

Fick's Law for Mass Diffusion Through Solids or Stagnant Fluids (Stationary Media)

Fick's law can be simplified for cases of dilute mass diffusion in solids, stagnant liquids, or stagnant gases. In these cases, $\rho_B \ll \rho$ and $v \approx 0$, which yields the following approximate result:

$$\dot{m}_B'' = J_B = -D_v \frac{d\rho_B}{dy} \tag{4}$$

part I *Theory*

Fick's Law for Ideal Gases with Negligible Temperature Gradient

For cases of dilute mass diffusion, Fick's law can be written in terms of partial pressure gradient instead of concentration gradient. When gas B can be approximated as ideal,

$$p_B = \frac{\rho_B R_u T}{M_B} = C_B R_u T \tag{5}$$

and when the gradient in T is small, Equation (3) can be written as

$$J_B = -\left(\frac{M_B D_v}{R_u T}\right)\frac{dp_B}{dy} \tag{6a}$$

or

$$J_B^* = -\left(\frac{D_v}{R_u T}\right)\frac{dp_B}{dy} \tag{6b}$$

If $v \approx 0$, Equation (4) may be written as

$$\dot{m}_B'' = J_B = -\left(\frac{M_B D_v}{R_u T}\right)\frac{dp_B}{dy} \tag{7a}$$

or

$$\dot{m}_B''^* = J_B^* = -\left(\frac{D_v}{R_u T}\right)\frac{dp_B}{dy} \tag{7b}$$

The partial pressure gradient formulation for mass transfer analysis has been used extensively; this is unfortunate because the pressure formulation [Equations (6) and (7)] applies only to cases where one component is dilute, the fluid closely approximates an ideal gas, and the temperature gradient has a negligible effect. The density (or concentration) gradient formulation expressed in Equations (1) through (4) is more general and can be applied to a wider range of mass transfer problems, including cases where neither component is dilute [Equation (1)]. The gases need not be ideal, nor the temperature gradient negligible. Consequently, this chapter emphasizes the density formulation.

Diffusion Coefficient

For a binary mixture, the diffusion coefficient D_v is a function of temperature, pressure, and composition. Experimental measurements of D_v for most binary mixtures are limited in range and accuracy. Table 1 gives a few experimental values for diffusion of some gases in air. For more detailed tables, see the section on Bibliography at the end of this chapter.

Table 1 Mass Diffusivities for Gases in Air[a]

Gas	D_v, mm²/s	Gas	D_v, mm²/s
Ammonia	27.9	Hydrogen	41.3
Benzene	8.8	Oxygen	20.6
Carbon dioxide	16.5	Water vapor	25.5
Ethanol	11.9		

[a] Gases at 25℃ and 101.325 kPa.

In the absence of data, use equations developed from (1) theory or (2) theory with constants adjusted from limited experimental data. For binary gas mixtures at low pressure, D_v is inversely proportional to pressure, increases with increasing temperature, and is almost independent of composition for a given gas pair. Bird et al. (1960) present the following equation, developed from kinetic theory and corresponding states arguments, for estimating D_v at pressures less than $0.1 p_{c\,min}$:

$$D_v = a \left(\frac{T}{\sqrt{T_{cA} + T_{cB}}} \right)^b \sqrt{\frac{1}{M_A} + \frac{1}{M_B}} \times \frac{(p_{cA} p_{cB})^{1/3} (T_{cA} T_{cB})^{5/12}}{p} \tag{8}$$

where

D_v = diffusion coefficient, mm²/s

a = constant, dimensionless

b = constant, dimensionless

T = absolute temperature, K

p = pressure, kPa

M = relative molecular mass, kg/kg mol

The subscripts cA and cB refer to the critical states of the two gases. Analysis of experimental data gives the following values of the constants a and b:

For nonpolar gas pairs,

$$a = 0.1280 \text{ and } b = 1.823$$

For water vapor with a nonpolar gas,

$$a = 0.1697 \text{ and } b = 2.334$$

A **nonpolar gas** is one for which the intermolecular forces are independent of the relative orientation of molecules, depending only on the separation distance from each other. Air, composed of nonpolar gases O_2 and N_2, is nonpolar.

Equation (8) is stated to agree with experimental data at atmospheric pressure to within about 8% (Bird et al. 1960).

The mass diffusivity D_v for binary mixtures at low pressure is predictable within about 10% by kinetic theory (Reid et al. 1987).

$$D_v = 0.5320 \frac{T^{1.5}}{p(\sigma_{AB})^2 \Omega_{D,AB}} \sqrt{\frac{1}{M_A} + \frac{1}{M_B}} \tag{9}$$

where

σ_{AB} = characteristic molecular diameter, nm

$\Omega_{D,AB}$ = temperature function, dimensionless

D_v is in mm²/s, p in kPa, and T in kelvins. If the gas molecules of A and B are considered rigid spheres having diameters σ_A and σ_B [and $\sigma_{AB} = (\sigma_A/2) + (\sigma_B/2)$], all expressed in nanometres, the dimensionless function $\Omega_{D,AB}$ equals unity. More realistic models for the molecules having intermolecular forces of attraction and repulsion lead to values of $\Omega_{D,AB}$ that are functions of temperature. Reid et al. (1987) present tabulations of this quantity. These results show that D_v increases as the 2.0 power of T at low temperatures and as the 1.65 power of T at very high temperatures.

The diffusion coefficient of moist air has been calculated for Equation (8) using a simplified intermolecular potential field function for water vapor and air (Mason and Monchick 1965).

The following is an empirical equation for mass diffusivity of water vapor in air up to 1100 ℃ (Sherwood and Pigford 1952):

$$D_v = \frac{0.926}{p} \left(\frac{T^{2.5}}{T+245} \right) \tag{10}$$

where D_v is in mm²/s, p in kPa, and T in kelvins.

Diffusion of One Gas Through a Second Stagnant Gas

Figure 1 shows diffusion of one gas through a second stagnant gas. Water vapor diffuses from the liquid surface into surrounding stationary air. It is assumed that local equilibrium exists through the gas mixture, that the gases are ideal, and that the Gibbs-Dalton law is valid, which implies that the temperature gradient has a negligible effect. Diffusion of water vapor is due to concentration gradient and is given by Equation (6a). There is a continuous gas phase, so the mixture pressure p is constant, and the Gibbs-Dalton law yields

$$p_A + p_B = p = \text{constant} \tag{11a}$$

or

$$\frac{\rho_A}{M_A} + \frac{\rho_B}{M_B} = \frac{\rho}{R_u T} = \text{constant} \tag{11b}$$

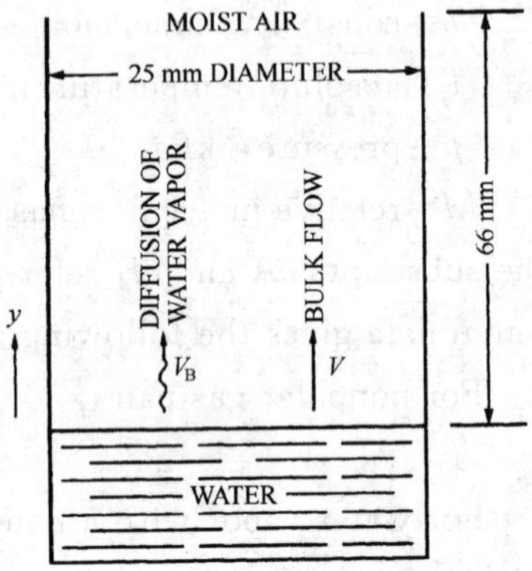

Fig. 1 Diffusion of Water Vapor Through Stagnant Air

The partial pressure gradient of the water vapor causes a partial pressure gradient of the air such that

$$\frac{dp_A}{dy} = -\frac{dp_B}{dy}$$

or

$$\left(\frac{1}{M_A}\right)\frac{dp_A}{dy} = -\left(\frac{1}{M_B}\right)\frac{dp_B}{dy} \tag{12}$$

Air, then, diffuses toward the liquid water interface. Because it cannot be absorbed there, a bulk velocity v of the gas mixture is established in a direction away from the liquid surface, so that the net transport of air is zero (i.e., the air is stagnant):

$$\dot{m}_A'' = -D_v \frac{dp_A}{dy} + \rho_A v = 0 \tag{13}$$

The bulk velocity v transports not only air but also water vapor away from the interface. Therefore, the total rate of water vapor diffusion is

$$\dot{m}_B'' = -D_v \frac{d\rho_B}{dy} + \rho_B v = 0 \tag{14}$$

Substituting for the velocity v from Equation (13) and using Equations (11b) and

(12) gives

$$\dot{m}_B'' = \left(\frac{D_v M_B p}{\rho_A R_u T}\right)\frac{d\rho_A}{dy} \quad (15)$$

Integration yields

$$\dot{m}_B'' = \frac{D_v M_B p}{R_u T}\left[\frac{\ln(\rho_{AL}/\rho_{A0})}{y_L - y_0}\right] \quad (16a)$$

or

$$\dot{m}_B'' = -D_v P_{Am}\left(\frac{\rho_{BL} - \rho_{B0}}{y_L - y_0}\right) \quad (16b)$$

where

$$P_{Am} = \frac{p}{p_{AL}}\rho_{AL}\left[\frac{\ln(\rho_{AL}/\rho_{A0})}{\rho_{AL} - \rho_{A0}}\right] \quad (17)$$

P_{Am} is the logarithmic mean density factor of the stagnant air. The pressure distribution for this type of diffusion is illustrated in Figure 2. **Stagnant** refers to the net behavior of the air; it does not move because the bulk flow exactly offsets diffusion. The term P_{Am} in Equation (16b) approximately equals unity for dilute mixtures such as water vapor in air at near atmospheric conditions. This condition makes it possible to simplify Equations (16) and implies that in the case of dilute mixtures, the partial pressure distribution curves in Figure 2 are straight lines.

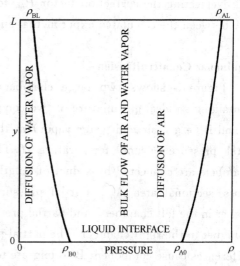

Fig. 2 Pressure Profiles for Diffusion of Water Vapor Through Stagnant Air

Example 1. A vertical tube of 25 mm diameter is partially filled with water so that the distance from the water surface to the open end of the tube is 60 mm, as shown in Figure 1. Perfectly dried air is blown over the open tube end, and the complete system is at a constant temperature of 15℃. In 200 h of steady operation, 2.15 g of water evaporates from the tube. The total pressure of the system is 101.325 kPa. Using these data, (a) calculate the mass diffusivity of water vapor in air, and (b) compare this experimental result with that from Equation (10).

Solution:

(a) The mass diffusion flux of water vapor from the water surface is

$$\dot{m}_B = 2.15/200 = 0.01075 \text{ g/h}$$

The cross-sectional area of a 25 mm diameter tube is $\pi(12.5)^2 = 491 \text{ mm}^2$. Therefore, $\dot{m}_B'' = 0.00608 \text{ g/(m}^2 \cdot \text{s)}$. The partial densities are determined with the aid of the psychrometric tables.

$$\rho_{BL} = 0; \quad \rho_{B0} = 12.8 \text{ g/m}^3$$

$$\rho_{AL} = 1.225 \text{ kg/m}^3; \quad \rho_{A0} = 1.204 \text{ kg/m}^3$$

Because $p = p_{AL} = 101.325$ kPa, the logarithmic mean density factor [Equation (17)] is

$$P_{Am} = 1.225 \left[\frac{\ln(1.225/1.204)}{1.225 - 1.204} \right] = 1.009$$

The mass diffusivity is now computed from Equation (16b) as

$$D_v = \frac{-\dot{m}''_B(y_L - y_0)}{P_{Am}(\rho_{BL} - \rho_{B0})} = \frac{-(0.00608)(0.060)(10^6)}{(1.009)(0 - 12.8)} = 28.2 \text{ mm}^2/\text{s}$$

(b) By Equation (10), with $p = 101.325$ kPa and $T = 15 + 273 = 288$ K,

$$D_v = \frac{0.926}{101.325} \left(\frac{288^{2.5}}{288 + 245} \right) = 24.1 \text{ mm}^2/\text{s}$$

Neglecting the correction factor P_{Am} for this example gives a difference of less than 1% between the calculated experimental and empirically predicted values of D_v.

Equimolar Counterdiffusion

Figure 3 shows two large chambers, both containing an ideal gas mixture of two components A and B (e.g., air and water vapor) at the same total pressure p and temperature T. The two chambers are connected by a duct of length L and cross-sectional area A_{cs}. Partial pressure p_B is higher in the left chamber, and partial pressure p_A is higher in the right chamber. The partial pressure differences cause component B to migrate to the right and component A to migrate to the left.

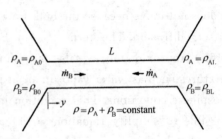

Fig. 3   Equimolar Counterdiffusion

At steady state, the molar flows of A and B must be equal, but in the opposite direction, or

$$\dot{m}''^*_A + \dot{m}''^*_B = 0 \tag{18}$$

because the total molar concentration C must stay the same in both chambers if p and T remain constant. Since the molar fluxes are the same in both directions, the molar average velocity $v^* = 0$. Thus, Equation (7b) can be used to calculate the molar flux of B (or A):

$$\dot{m}''^*_B = \frac{-D_v}{R_u T} \frac{dp_B}{dy} \tag{19}$$

or

$$\dot{m}^*_B = \frac{A_{cs} D_v}{R_u T} \frac{p_{B0} - p_{BL}}{L} \tag{20}$$

or

$$\dot{m}_B = \frac{M_B A_{cs} D_v}{R_u T} \frac{p_{B0} - p_{BL}}{L} \tag{21}$$

Example 2. One large room is maintained at 22°C (295K), 101.3kPa, 80% rh. A 20m long duct with cross-sectional area of 0.15m² connects the room to another large room at 22°C, 101.3kPa, 10% rh. What is the rate of water vapor diffusion between the two rooms?

Solution: Let air be component A and water vapor be component B.

Equation (21) can be used to calculate the mass flow of water vapor B.

Equation (10) can be used to calculate the diffusivity.

$$D_v = \frac{0.926}{101.3}\left(\frac{295^{2.5}}{295+245}\right) = 25.3 \text{mm}^2/\text{s}$$

From a psychrometric table, the saturated vapor pressure at 22℃ is 2.645kPa. The vapor pressure difference $p_{B0} \cdot p_{BL}$ is

$$p_{B0} - p_{BL} = (0.8-0.1)2.645\text{kPa} = 1.85\text{kPa}$$

Then, Equation (21) gives

$$\dot{m}_B = \frac{18 \times 0.15(25.3/10^6)}{8.314 \times 295}\frac{1.85}{20} = 2.58 \times 10^{-9} \text{kg/s}$$

Molecular Diffusion in Liquids and Solids

Because of the greater density, diffusion is slower in liquids than in gases. No satisfactory molecular theories have been developed for calculating diffusion coefficients. The limited measured values of D_v show that, unlike for gas mixtures at low pressures, the diffusion coefficient for liquids varies appreciably with concentration.

Reasoning largely from analogy to the case of one-dimensional diffusion in gases and employing Fick's law as expressed by Equation (4) gives

$$\dot{m}''_B = D_v\left(\frac{\rho_{B1} - \rho_{B2}}{y_2 - y_1}\right) \tag{22}$$

Equation (22) expresses the steady-state diffusion of the solute B through the solvent A in terms of the molal concentration difference of the solute at two locations separated by the distance $\Delta y = y_2 - y_1$. Bird et al. (1960), Hirschfelder et al. (1954), Sherwood and Pigford (1952), Reid and Sherwood (1966), Treybal (1980), and Eckert and Drake (1972) provide equations and tables for evaluating D_v. Hirschfelder et al. (1954) provide comprehensive treatment of the molecular developments.

Diffusion through a solid when the solute is dissolved to form a homogeneous solid solution is known as **structure-insensitive diffusion** (Treybal 1980). This solid diffusion closely parallels diffusion through fluids, and Equation (22) can be applied to one-dimensional steady-state problems. Values of mass diffusivity are generally lower than they are for liquids and vary with temperature.

The diffusion of a gas mixture through a porous medium is common (e.g., the diffusion of an air-vapor mixture through porous insulation). The vapor diffuses through the air along the tortuous narrow passages within the porous medium. The mass flux is a function of the vapor pressure gradient and diffusivity as indicated in Equation (7a). It is also a function of the structure of the pathways within the porous medium and is therefore called **structure-sensitive diffusion**. All of these factors are taken into account in the following version of Equation (7a):

$$\dot{m}''_B = -\bar{\mu}\frac{dp_B}{dy} \tag{23}$$

where $\bar{\mu}$ is called the permeability of the porous medium.

CONVECTION OF MASS

Convection of mass involves the mass transfer mechanisms of molecular diffusion and bulk fluid motion. Fluid motion in the region adjacent to a mass transfer surface may be laminar or turbulent, depending on geometry and flow conditions.

Mass Transfer Coefficient

Convective mass transfer is analogous to convective heat transfer where geometry and boundary conditions are similar. The analogy holds for both laminar and turbulent flows and applies to both external and internal flow problems.

Mass Transfer Coefficients for External Flows. Most external convective mass transfer problems can be solved with an appropriate formulation that relates the mass transfer flux (to or from an interfacial surface) to the concentration difference across the boundary layer illustrated in Figure 4. This formulation gives rise to the convective mass transfer coefficient, defined as

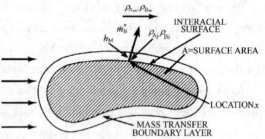

Fig. 4 Nomenclature for Convective Mass Transfer from External Surface at Location x Where Surface is Impermeable to Gas A

$$h_M = \frac{\dot{m}_B''}{\rho_{Bi} - \rho_{B\infty}} \tag{24}$$

where

h_M = local external mass transfer coefficient, m/s

$\dot{m}_B''$ = mass flux of gas B from surface, kg/(m² · s)

ρ_{Bi} = density of gas B at interface (saturation density), kg/m³

$\rho_{B\infty}$ = density of component B outside boundary layer, kg/m³

If ρ_{Bi} and $\rho_{B\infty}$ are constant over the entire interfacial surface, the mass transfer rate from the surface can be expressed as

$$\dot{m}_B'' = \bar{h}_M (\rho_{Bi} - \rho_{B\infty}) \tag{25}$$

where $\bar{h}_M$ is the average mass transfer coefficient, defined as

$$\bar{h}_M \equiv \frac{1}{A} \int_A h_m \, dA \tag{26}$$

Mass Transfer Coefficients for Internal Flows. Most internal convective mass transfer problems, such as those that occur in channels or in the cores of dehumidification coils, can be solved if an appropriate expression is available to relate the mass transfer flux (to or from the interfacial surface) to the difference between the concentration at the surface and the bulk concentration in the channel, as illustrated in Figure 5. This formulation leads to the definition of the mass transfer coefficient for internal flows:

$$h_M \equiv \frac{\dot{m}_B''}{\rho_{Bi} - \rho_{Bb}} \quad (27)$$

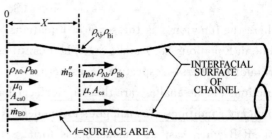

Fig. 5 Nomenclature for Convective Mass Transfer from Internal Surface Impermeable to Gas A

where

h_M = internal mass transfer coefficient, m/s

$\dot{m}_B''$ = mass flux of gas B at interfacial surface, kg/(m² · s)

ρ_{Bi} = density of gas B at interfacial surface, kg/m³

$\rho_{Bb} \equiv (1/\bar{u}_B A_{cs}) \int_{A_{cs}} u_B \rho_B dA_{cs}$ = bulk density of gas B at location x

$\bar{u}_B \equiv (1/A_{cs}) \int_A u_B dA_{cs}$ = average velocity of gas B at location x, m/s

A_{cs} = cross-sectional area of channel at station x, m²

u_B = velocity of component B in x direction, m/s

ρ_B = density distribution of component B at station x, kg/m³

Often, it is easier to obtain the bulk density of gas B from

$$\rho_{Bb} = \frac{\dot{m}_{B0} + \int_A \dot{m}_B'' dA}{\bar{u}_B A_{cs}} \quad (28)$$

where

$\dot{m}_{B0}$ = mass flow rate of component B at station $x=0$, kg/s

A = interfacial area of channel between station $x=0$ and station $x=x$, m²

Equation (28) can be derived from the preceding definitions. The major problem is the determination of $\bar{u}_B$. If, however, the analysis is restricted to cases where B is dilute and concentration gradients of B in the x direction are negligibly small, $\bar{u}_B \approx \bar{u}$. Component B is swept along in the x direction with an average velocity equal to the average velocity of the dilute mixture.

Analogy Between Convective Heat and Mass Transfer

Most expressions for the convective mass transfer coefficient h_M are determined from expressions for the convective heat transfer coefficient h.

For problems in internal and external flow where mass transfer occurs at the convective surface and where component B is dilute, it is shown by Bird et al. (1960) and Incropera and DeWitt (1996) that the Nusselt and Sherwood numbers are defined as follows:

$$\text{Nu} = f(X, Y, Z, \text{Pr}, \text{Re}) \quad (29)$$
$$\text{Sh} = f(X, Y, Z, \text{Sc}, \text{Re}) \quad (30)$$

and
$$\overline{\text{Nu}} = g(\text{Pr}, \text{Re}) \quad (31)$$

$$\overline{Sh} = g(Sc, Re) \tag{32}$$

where the function f is the same in Equations (29) and (30), and the function g is the same in Equations (31) and (32). The quantities Pr and Sc are dimensionless Prandtl and Schmidt numbers, respectively, as defined in the section on Symbols. The primary restrictions on the analogy are that the surface shapes are the same and that the temperature boundary conditions are analogous to the density distribution boundary conditions for component B when cast in dimensionless form. Several primary factors prevent the analogy from being perfect. In some cases, the Nusselt number was derived for smooth surfaces. Many mass transfer problems involve wavy, droplet-like, or roughened surfaces. Many Nusselt number relations are obtained for constant temperature surfaces. Sometimes ρ_{Bi} is not constant over the entire surface because of varying saturation conditions and the possibility of surface dryout.

In all mass transfer problems, there is some blowing or suction at the surface because of the condensation, evaporation, or transpiration of component B. In most cases, this blowing/suction phenomenon has little effect on the Sherwood number, but the analogy should be examined closely if $v_i/u_\infty > 0.01$ or $v_i/\overline{u} > 0.01$, especially if the Reynolds number is large.

Example 3. Use the analogy expressed in Equations (31) and (32) to solve the following problem. An expression for heat transfer from a constant temperature flat plate in laminar flow is

$$\overline{Nu}_L = 0.664 Pr^{1/3} Re_L^{1/2} \tag{33}$$

$Sc = 0.35$, $D_v = 3.6 \times 10^{-5}$ m²/s, and $Pr = 0.708$ for the given conditions; determine the mass transfer rate and temperature of the waterwetted flat plate in Figure 6 using the heat/mass transfer analogy.

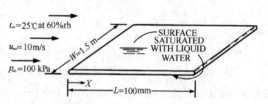

Fig. 6 Water-Saturated Flat Plate in Flowing Airstream

Solution: To solve the problem, properties should be evaluated at film conditions. However, since the plate temperature and the interfacial water vapor density are not known, a first estimate will be obtained assuming the plate t_{i1} to be at 25°C. The plate Reynolds number is

$$Re_{L1} = \frac{\rho u_\infty L}{\mu} = \frac{(1.166 \text{kg/m}^3)(10\text{m/s})(0.1\text{m})}{[1.965 \times 10^{-5} \text{kg/(m} \cdot \text{s)}]} = 59340$$

The plate is entirely in laminar flow, since the transitional Reynolds number is about 5×10^5. Using the mass transfer analogy given by Equations (31) and (32), Equation (33) yields

$$\overline{Sh}_{L1} = 0.664 Sc^{1/3} Re_L^{1/2} = 0.664(0.35)^{1/3}(59340)^{1/2} = 114$$

From the definition of the Sherwood number,

$$\overline{h}_{M1} = \overline{Sh}_{L1} D_v/L = (114)(3.6 \times 10^{-5} \text{m}^2/\text{s})/(0.1\text{m}) = 0.04104 \text{m/s}$$

The psychrometric tables give a humidity ratio W of 0.0121 at 25°C and 60%

rh. Therefore,
$$\rho_{B\infty} = 0.0121\rho_{A\infty} = (0.0121)(1.166\text{kg/m}^3) = 0.01411\text{kg/m}^3$$
From steam tables, the saturation density for water at 25℃ is
$$\rho_{Bi} = 0.02352\text{kg/m}^3$$
Therefore, the mass transfer rate from the double-sided plate is
$$\dot{m}_{B1} = \bar{h}_{M1}A(\rho_{Bi} - \rho_{B\infty})$$
$$= (0.04104\text{m/s})(0.1\text{m} \times 1.5\text{m} \times 2) \times (0.02352\text{kg/m}^3 - 0.01411\text{kg/m}^3)$$
$$= 1.159 \times 10^{-4}\text{kg/s} = 0.1159\text{g/s}$$

This mass rate, transformed from the liquid state to the vapor state, requires the following heat rate to the plate to maintain the evaporation:
$$q_{i1} = \dot{m}_{B1}h_{fg} = (0.1159\text{g/s})(2443\text{kJ/kg}) = 283.1\text{W}$$

To obtain a second estimate of the wetted plate temperature in this type of problem, the following criteria are used. Calculate the t_i necessary to provide a heat rate of q_{i1}. If this temperature t_{iq1} is above the dew-point temperature t_{id}, set the second estimate at $t_{i2} = (t_{iq1} + t_{i1})/2$. If t_{iq1} is below the dew-point temperature, set $t_{i2} = (t_{id} + t_{i1})/2$. For this problem, the dew point is $t_{id} = 14℃$.

Obtaining the second estimate of the plate temperature requires an approximate value of the heat transfer coefficient.
$$\overline{\text{Nu}}_{L1} = 0.664\text{Pr}^{1/3}\text{Re}_L^{1/2} = 0.664(0.708)^{1/3}(59340)^{1/2} = 144.2$$
From the definition of the Nusselt number,
$$\bar{h}_1 = \overline{\text{Nu}}_{L1}k/L = (144.2)[0.0261\text{W/(m·K)}]/0.1\text{m} = 37.6\text{W/(m}^2\text{·K)}$$
Therefore, the second estimate for the plate temperature is
$$t_{iq1} = t_\infty - q_{i1}/(\bar{h}_1 A) = 25℃ - \{283.1\text{W}/[37.6\text{W/(m}^2\text{·K)} \times (2 \times 0.1\text{m} \times 1.5\text{m})]\}$$
$$= 25℃ - 25℃ = 0℃$$
This temperature is below the dew-point temperature, therefore,
$$t_{i2} = (14℃ + 25℃)/2 = 19.5℃$$
The second estimate of the film temperature is
$$t_{f2} = (t_{i2} + t_\infty)/2 = (19.5℃ + 25℃)/2 = 22.25℃$$
The next iteration on the solution is as follows:
$$\text{Re}_{L2} = 61010$$
$$\overline{\text{Sh}}_{L2} = 0.664(0.393)^{1/3}(61010)^{1/2} = 120$$
$$\bar{h}_{M2} = (120)(3.38 \times 10^{-5})/0.1 = 0.04056\text{m/s}$$

The free stream density of the water vapor has been evaluated. The density of the water vapor at the plate surface is the saturation density at 19.5℃.
$$p_{Bi2} = (0.01374)(1.183\text{kg/m}^3) = 0.01625\text{kg/m}^3$$
$$A = 2 \times 0.1 \times 1.5 = 0.3\text{m}^2$$
$$\dot{m}_{B2} = (0.04056\text{m/s})(0.3\text{m}^2)(0.01625\text{kg/m}^3 - 0.01411\text{kg/m}^3)$$
$$= 2.604 \times 10^{-5}\text{kg/s} = 0.02604\text{g/s}$$

$$q_{i2} = (0.02604 \text{g/s})(2458 \text{J/kg}) = 64.01 \text{W}$$

$$\overline{\text{Nu}}_{L2} = 0.664(0.709)^{1/3}(61010)^{1/2} = 146$$

$$\overline{h}_2 = (146)(0.02584)/0.1 = 37.73 \text{W}(\text{m}^2 \cdot \text{K})$$

$$t_{iq2} = 25\text{°C} - [(64.01\text{W})/(37.73 \times 0.3)] = 19.34\text{°C}$$

This temperature is above the dew-point temperature; therefore,

$$t_{i3} = (t_{i2} + t_{iq2})/2 = (19.5 + 19.34)/2 = 19.42\text{°C}$$

This is approximately the same result as that obtained in the previous iteration. Therefore, the problem solution is

$$t_i = 19.5\text{°C}$$

$$\dot{m}_B = 0.0260 \text{g/s}$$

The kind of similarity between heat and mass transfer that results in Equation (29) through Equation (32) can also be shown to exist between heat and momentum transfer. Chilton and Colburn (1934) used this similarity to relate Nusselt number to friction factor by the analogy

$$j_H = \frac{\text{Nu}}{\text{Re Pr}^{(1-n)}} = \text{St Pr}^n = \frac{f}{2} \quad (34)$$

where $n = 2/3$, $\text{St} = \text{Nu}/(\text{Re Pr})$ is the Stanton number, and j_H is the Chilton-Colburn j-factor for heat transfer. Substituting Sh for Nu and Sc for Pr in Equations (31) and (32) gives the Chilton-Colburn j-factor for mass transfer, j_D:

$$j_D = \frac{\text{Sh}}{\text{Re Sc}^{(1-n)}} = \text{St}_m \text{Sc}^n = \frac{f}{2} \quad (35)$$

where $\text{St}_m = \text{Sh}P_{AM}/(\text{Re Sc})$ is the Stanton number for mass transfer. Equations (34) and (35) are called the **Chilton-Colburn j-factor analogy.**

The power of the Chilton-Colburn j-factor analogy is represented in Figure 7 through 10. Figure 7 plots various experimental values of j_D from a flat plate with flow parallel to the plate surface. The solid line, which represents the data to near perfection, is actually $f/2$ from Blasius' solution of laminar flow on a flat plate (left-hand portion of the solid line) and Goldstein's solution for a turbulent boundary layer (right-hand portion). The right-hand portion of the solid line also represents McAdams' (1954) correlation of turbulent flow heat transfer coefficient for a flat plate.

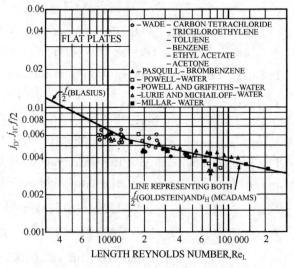

Fig. 7 Mass Transfer from Flat Plate

A wetted-wall column is a vertical tube in which a thin liquid film adheres to the tube surface and exchanges mass by evaporation or absorption with a gas flowing through the tube. Figure 8 illustrates typical data on vaporization in wetted-wall columns, plotted as j_D versus Re. The spread of the points with variation in $\mu/\rho D_v$ results from Gilliland's finding of an exponent of 0.56, not 2/3, representing the effect of the Schmidt number. Gilliland's equation can be written as follows:

$$j_D = 0.023 \mathrm{Re}^{-0.17} \left(\frac{\mu}{\rho D_v}\right)^{-0.56} \quad (36)$$

Similarly, McAdams' (1954) equation for heat transfer in pipes can be expressed as

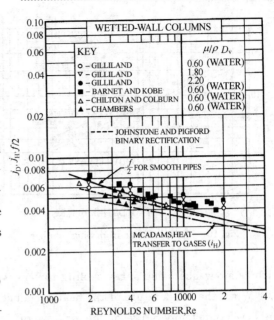

Fig. 8 Vaporization and Absorption in Wetted-Wall Column

$$j_H = 0.023 \mathrm{Re}^{-0.20} \left(\frac{c_p \mu}{k}\right)^{-0.7} \quad (37)$$

This is represented by the dash-dot curve in Figure 8, which falls below the mass transfer data. The curve $f/2$ representing friction in smooth tubes is the upper, solid curve.

Data for the evaporation of liquids from single cylinders into gas streams flowing transversely to the cylinders' axes are shown in Figure 9. Although the dash-dot line on Figure 9 represents the data, it is actually taken from McAdams (1954) as representative of a large collection of data on heat transfer to single cylinders placed transverse to airstreams. To compare these data with friction, it is necessary to distinguish between total drag and skin friction. Since the analogies are based on skin friction, the normal pressure drag must be subtracted from the measured total drag. At $\mathrm{Re} = 1000$, the skin friction is 12.6% of the total drag; at $\mathrm{Re} = 31600$, it is only 1.9%. Consequently, the values of $f/2$ at a high Reynolds number, obtained by the difference, are subject to considerable error.

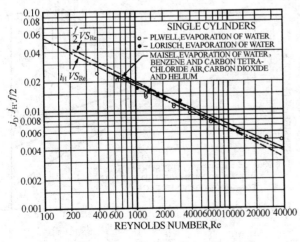

Fig. 9 Mass Transfer from Single Cylinders in Crossflow

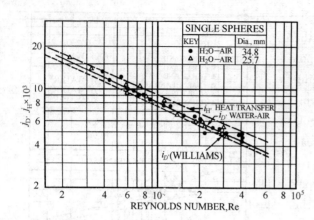

Fig. 10 Mass Transfer from Single Spheres

In Figure 10, data on the evaporation of water into air for single spheres are presented. The solid line, which best represents these data, agrees with the dashed line representing McAdams' correlation for heat transfer to spheres. These results cannot be compared with friction or momentum transfer because total drag has not been allocated to skin friction and normal pressure drag. Application of these data to air-water contacting devices such as air washers and spray cooling towers is well substantiated.

When the temperature of the heat exchanger surface in contact with moist air is below the dew-point temperature of the air, vapor condensation occurs. Typically, the air dry-bulb temperature and humidity ratio both decrease as the air flows through the exchanger. Therefore, sensible and latent heat transfer occur simultaneously. This process is similar to one that occurs in a spray dehumidifier and can be analyzed using the same procedure; however, this is not generally done.

Cooling coil analysis and design are complicated by the problem of determining transport coefficients h, h_M, and f. It would be convenient if heat transfer and friction data for dry heating coils could be used with the Colburn analogy to obtain the mass transfer coefficients. However, this approach is not always reliable, and work by Guillory and McQuiston (1973) and Helmer (1974) shows that the analogy is not consistently true. Figure 11 shows j-factors for a simple parallel plate exchanger for different surface conditions with sensible heat transfer. Mass transfer j-factors and the friction factors exhibit the same behavior. Dry surface j-factors fall below those obtained under dehumidifying conditions with the surface wet. At low Reynolds numbers, the boundary layer grows quickly; the droplets are soon covered and have little effect on the flow field. As the Reynolds number is increased, the boundary layer becomes thin and more of the total flow field is exposed to the droplets. The roughness caused by the droplets induces mixing and larger j-

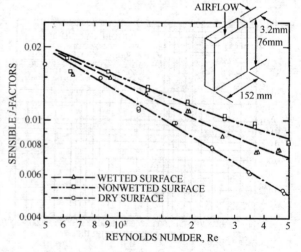

Fig. 11 Sensible Heat Transfer j-Factors for Parallel Plate Exchanger

factors. The data in Figure 11 cannot be applied to all surfaces because the length of the flow channel is also an important variable. However, the water collecting on the surface is mainly responsible for breakdown of the j-factor analogy. The j-factor analogy is approximately true when the surface conditions are identical. Under some conditions, it is possible to obtain a film of condensate on the surface instead of droplets. Guillory and McQuiston (1973) and Helmer (1974) related dry sensible j-and f-factors to those for wetted dehumidifying surfaces.

The equality of j_H, j_D, and $f/2$ for certain streamline shapes at low mass transfer rates has experimental verification. For flow past bluff objects, j_H and j_D are much smaller than $f/2$, based on total pressure drag. The heat and mass transfer, however, still relate in a useful way by equating j_H and j_D.

Example 4. Using solid cylinders of volatile solids (e. g., naphthalene, camphor, dichlorobenzene) with airflow normal to these cylinders, Bedingfield and Drew (1950) found that the ratio between the heat and mass transfer coefficients could be closely correlated by the following relation:

$$\frac{h}{\rho h_M} = [1230 \text{J}/(\text{kg} \cdot \text{K})] \left(\frac{\mu}{\rho D_v}\right)^{0.56}$$

For completely dry air at 21℃ flowing at a velocity of 9.5 m/s over a wet-bulb thermometer of diameter $d=7.5$ mm, determine the heat and mass transfer coefficients from Figure 9 and compare their ratio with the Bedingfield-Drew relation.

Solution: For dry air at 21℃ and standard pressure, $\rho=1.198 \text{kg/m}^3$, $\mu=1.82\times10^{-5}$ kg/(s·m), $k=0.02581$ W/(m·K), and $c_p=1.006$ kJ/(kg·K). From Equation (10), $D_v=25.13 \text{mm}^2/\text{s}$. Therefore,

$$\text{Re}_{da} = \rho u_\infty d/\mu = 1.198 \times 9.5 \times 7.5/(1000 \times 1.82 \times 10^{-5}) = 4690$$

$$\text{Pr} = c_p \mu/k = 1.006 \times 1.82 \times 10^{-5} \times 1000/(0.02581) = 0.709$$

$$\text{Sc} = \mu/\rho D_v = 1.82 \times 10^{-5} \times 10^6/(1.198 \times 25.13) = 0.605$$

From Figure 9 at $\text{Re}_{da}=4700$, read $j_H=0.0089$, $j_D=0.010$. From Equations (34) and (35),

$$h = j_H \rho c_p u_\infty/(\text{Pr})^{2/3} = 0.0089 \times 1.198 \times 1.006 \times 9.5 \times 1000/(0.709)^{2/3} = 238 \text{W}/(\text{m}^2 \cdot \text{K})$$

$$h_M = j_D u_\infty/(\text{Sc})^{2/3} = 0.010 \times 9.5/(0.605)^{2/3} = 0.133 \text{m/s}$$

$$h/\rho h_M = 128/(1.198 \times 0.133) = 803 \text{J}/(\text{kg} \cdot \text{K})$$

From the Bedingfield-Drew relation,

$$h/\rho h_M = 1230(0.605)^{0.56} = 928 \text{J}/(\text{kg} \cdot \text{K})$$

Equations (34) and (35) are call the Reynolds analogy when $\text{Pr}=\text{Sc}=1$. This suggests that $h/\rho h_M = c_p = 1006 \text{J}/(\text{kg} \cdot \text{K})$. This close agreement is because the ratio Sc/Pr is 0.605/0.709 or 0.85, so that the exponent of these numbers has little effect on the ratio of the transfer coefficients.

The extensive developments for calculating heat transfer coefficients can be applied to calculate mass transfer coefficients under similar geometrical and flow conditions using the

j-factor analogy. Similarly, mass transfer experiments often replace corresponding heat transfer experiments with complex geometries where exact boundary conditions are difficult to model (Sparrow and Ohadi 1987a, 1987b).

The j-factor analogy is useful only at low mass transfer rates. As the rate of mass transfer increases, the movement of matter normal to the transfer surface increases the convective velocity. For example, if a gas is blown from many small holes in a flat plate placed parallel to an airstream, the boundary layer thickens, and resistance to both mass and heat transfer increases with increasing blowing rate. Heat transfer data are usually collected at zero or, at least, insignificant mass transfer rates. Therefore, if such data are to be valid for a mass transfer process, the mass transfer rate (i.e., the blowing) must be low.

The j-factor relationship $j_H = j_D$ can still be valid at high mass transfer rates, but neither j_H nor j_D can be represented by data at zero mass transfer conditions.

Lewis Relation

Heat and mass transfer coefficients are satisfactorily related at the same Reynolds number by equating the Chilton-Colburn j-factors. Comparing Equations (34) and (35) gives

$$\text{St Pr}^n = \frac{f}{2} = \text{St}_m \text{Sc}^n$$

Inserting the definitions of St, Pr, St_m, and Sc gives

$$\frac{h}{\rho c_p \bar{u}} \left(\frac{c_p \mu}{k} \right)^{2/3} = \frac{h_M P_{Am}}{\bar{u}} \left(\frac{\mu}{\rho D_v} \right)^{2/3}$$

or

$$\frac{h}{h_M \rho c_p} = P_{Am} \left[\frac{(\mu/\rho D_v)}{(c_p \mu/k)} \right]^{2/3} = P_{Am} (\alpha/D_v)^{2/3} \qquad (38)$$

The quantity α/D_v is the **Lewis number** Le. Its magnitude expresses relative rates of propagation of energy and mass within a system. It is fairly insensitive to temperature variation. For air and water vapor mixtures, the ratio is (0.60/0.71) or 0.845, and $(0.845)^{2/3}$ is 0.894. At low diffusion rates, where the heat-mass transfer analogy is valid, P_{Am} is essentially unity. Therefore, for air and water vapor mixtures,

$$\frac{h}{h_M \rho c_p} \approx 1 \qquad (39)$$

The ratio of the heat transfer coefficient to the mass transfer coefficient is equal to the specific heat per unit volume of the mixture at constant pressure. This relation [Equation (39)] is usually called the Lewis relation and is nearly true for air and water vapor at low mass transfer rates. It is generally not true for other gas mixtures because the ratio Le of thermal to vapor diffusivity can differ from unity. The agreement between wet-bulb temperature and adiabatic saturation temperature is a direct result of the nearness of the Lewis number to unity for air and water vapor.

The Lewis relation is valid in turbulent flow whether or not α/D_v equals 1 because eddy diffusion in turbulent flow involves the same mixing action for heat exchange as for mass exchange, and this action overwhelms any molecular diffusion. Deviations from the Lewis relation are, therefore, due to a laminar boundary layer or a laminar sublayer and buffer zone where molecular transport phenomena are the controlling factors.

WORDS AND EXPRESSIONS

absorbent	*adj.* 能吸收的；*n.* 吸收剂
adhesive	*adj.* 带黏性的，胶粘；*n.* 粘合剂
air washer	空气洗涤器；净气器；净气器；净气装置；净气装置；淋水系统
air-side	空气侧
analogy	*n.* 类似，类推
bonding	*n.* 连［搭，焊，胶，粘］接，结［耦，焊，接］合，压焊
camphor	*n.* ［化］樟脑
capillary	*n.* 毛细管；*adj.* 毛状的，毛细作用的；
characteric length	特征长度
concentration	*n.* 集中，集合，专心，浓缩，浓度
corrugated	*adj.* 缩成皱纹的，使起波状的
cumbersome	*adj.* 讨厌的，麻烦的，笨重的
dash-dot curve	点划线
dehumidification	*n.* 除湿
dew-point	*n.* 露点
dichlorobenzene	*n.* ［化］二氯（代）苯
drainage	*n.* 排水，排泄，排水装置，排水区域，排出物，消耗
empirical equation	经验公式
enthalpy	*n.* ［物］焓，热函
epoxy resin	*n.* 环氧树脂
equilibrium	*n.* 平衡，平静，均衡，保持平衡的能力，沉着，安静
erroneous result	错误的结果
exponential	*n.* 指数，倡导者，演奏者，例子，［数］指数；*adj.* 指数的，幂数的
extrapolation	*n.* ［数］外推法，推断
extrude	*v.* 挤压出，挤压成，突出，伸出，逐出
Fanning friction factor	范宁摩擦系数
granular	*adj.* 由小粒而成的，粒状的
impermeable	*adj.* 不能渗透的，不渗透性的

part Ⅰ *Theory*

interpolate	*v.* 窜改，添写进去，插入新语句
interstice	*n.* 空隙，裂缝
latent heat	潜热
metallurgical	*adj.* 冶金学的
nanometer	*n.* 纳米
naphthalene	*n.* 萘(球)，卫生球
nucleate boiling	核态沸腾
offset	*n.* 偏移量，抵销，弥补，分支，平版印刷，胶印；*vt.* 弥补，抵销，用平版印刷；*vi.* 偏移，形成分支
overwhelm	*vt.* 淹没，覆没，受打击，制服，压倒
ozone	*n.* [化] 臭氧
plausible	*adj.* 似是而非的
psychrometer	*n.* 干湿球湿度计，干湿计
repulsion	*n.* 推斥，排斥，严拒，厌恶，反驳
sensible heat	显热
staggered	*adj.* 错列的，叉排的
stagnant	*adj.* 停滞的，迟钝的（stagnant fluid 静止流体）
tabular	*adj.* 制成表的，扁平的，表格式的，平坦的；*vi.* 列表，排成表格式
tapered	*adj.* 锥形的；渐缩的
transverse	*adj.* 横向的，横断的
volumetric flow rate	体积流量

NOTATIONS

(1) 注意下列词组和短语的用法：
Expressing（表示，表达）the concentration of component B of a binary mixture <u>in terms of</u>（根据，按照，用…的话，在…方面）the mass fraction ρ_B/ρ or mole fraction C_B/C, Fick's law is……

<u>For cases of</u>（对于……的情况）dilute mass diffusion, Fick's law can be written <u>in terms of</u>（同上）<u>pressure gradient</u>（压力梯度）instead of <u>concentration gradient</u>（浓度梯度）.

For binary gas mixtures at low pressure, D_v <u>is inversely proportional to</u>（与……成反比）pressure, <u>increases with increasing</u>（随……的增加而增加）temperature, and is almost independent of composition for a given gas pair.

rate of heat transfer(W), [传热速率（单位"W"）]
heat flux per unit area(W/m^2), [单位面积传热通量（单位"W/m^2"）]
velocity in x direction (x 方向上的速度)

(2) In the absence of data, use equations developed from (1) theory or (2) theory with constants adjusted from limited experimental data.

在缺乏数据的情况下，则使用由有限实验数据整理得到常数的(1)理论或(2)理论发展起来的公式。

(3) ...... where D_v is in mm^2/s, p in kPa, and T in kelvins.

其中，D_v 单位是 mm^2/s，p 单位是 kPa，and T 单位是 "K"。

EXERCISES

(1) 请把下列词语翻成中文，并熟练使用。

air washer, cooling coil, evaporative condenser, cooling tower

(2) 请翻译下列句子。

1) The primary restrictions on the analogy are that the surface shapes are the same and that the dimensionless temperature boundary conditions are analogous to the dimensionless density distribution boundary conditions for component B when cast in dimensionless form. Several primary factors prevent the analogy from being perfect. In some cases, the Nusselt number was derived for smooth surfaces. Many mass transfer problems involve wavy, droplet-like, or roughened surfaces. Many Nusselt number relations are obtained for constant temperature surfaces. Sometimes ρ_{Bi} is not constant over the entire surface because of varying saturation conditions and the possibility of surface dryout.

2) 当 $\rho_B < 0.10\rho$，温度低于 60℃时，如果 J_B 误差为 10% 是可容忍的，方程(3)则仍然适用。

LESSON 5

PSYCHROMETRICS

PSYCHROMETRICS deals with the thermodynamic properties of moist air and uses these properties to analyze conditions and processes involving moist air.

Hyland and Wexler (1983a, b) developed formulas for thermodynamic properties of moist air and water. However, perfect gas relations can be used instead of these formulas in most air-conditioning problems. Kuehn et al. (1998) showed that errors are less than 0.7% in calculating humidity ratio, enthalpy, and specific volume of saturated air at standard atmospheric pressure for a temperature range of −50 to 50℃. Furthermore, these errors decrease with decreasing pressure.

This chapter discusses perfect gas relations and describes their use in common air-conditioning problems. The formulas developed by Hyland and Wexler (1983a) and discussed by Olivieri (1996) may be used where greater precision is required.

COMPOSITION OF DRY AND MOIST AIR

Atmospheric air contains many gaseous components as well as water vapor and miscellaneous contaminants (e.g., smoke, pollen, and gaseous pollutants not normally present in free air far from pollution sources).

Dry air exists when all water vapor and contaminants have been removed from atmospheric air. The composition of dry air is relatively constant, but small variations in the amounts of individual components occur with time, geographic location, and altitude. Harrison (1965) lists the approximate percentage composition of dry air by volume as: nitrogen, 78.084; oxygen, 20.9476; argon, 0.934; carbon dioxide, 0.0314; neon, 0.001818; helium, 0.000524; methane, 0.00015; sulfur dioxide, 0 to 0.0001; hydrogen, 0.00005; and minor components such as krypton, xenon, and ozone, 0.0002. The relative molecular mass of all components for dry air is 28.9645, based on the carbon-12 scale (Harrison 1965). The gas constant for dry air, based on the carbon-12 scale, is

$$R_{da} = 8314.41/28.9645 = 287.055 \text{ J/(kg} \cdot \text{K)} \tag{1}$$

Moist air is a binary (two-component) mixture of dry air and water vapor. The amount of water vapor in moist air varies from zero (dry air) to a maximum that depends on temperature and pressure. The latter condition refers to **saturation**, a state of neutral equilibrium

---

Extracted from chaper 6 of the ASHREA *Handbook-FundaMentals*.

between moist air and the condensed water phase (liquid or solid). Unless otherwise stated, saturation refers to a flat interface surface between the moist air and the condensed phase. Saturation conditions will change when the interface radius is very small such as with ultrafine water droplets. The relative molecular mass of water is 18.01528 on the carbon-12 scale. The gas constant for water vapor is

$$R_w = 8314.41/18.01528 = 461.520 \, \text{J}/(\text{kg} \cdot \text{K}) \tag{2}$$

UNITED STATES STANDARD ATMOSPHERE

The temperature and barometric pressure of atmospheric air vary considerably with altitude as well as with local geographic and weather conditions. The standard atmosphere gives a standard of reference for estimating properties at various altitudes. At sea level, standard temperature is 15℃; standard barometric pressure is 101.325kPa. The temperature is assumed to decrease linearly with increasing altitude throughout the troposphere (lower atmosphere), and to be constant in the lower reaches of the stratosphere. The lower atmosphere is assumed to consist of dry air that behaves as a perfect gas. Gravity is also assumed constant at the standard value, 9.80665m/s². Table 1 summarizes property data for altitudes to 10000m.

Standard Atmospheric Data for Altitudes to 10000m   Table 1

Altitude, m	Temperature, ℃	Pressure, kPa	Altitude, m	Temperature, ℃	Pressure, kPa
−500	18.2	107.478	6000	−24.0	47.181
0	15.0	101.325	7000	−30.5	41.061
500	11.8	95.461	8000	−37.0	35.600
1000	8.5	89.875	9000	−43.5	30.742
1500	5.2	84.556	10000	−50	26.436
2000	2.0	79.495	12000	−63	19.284
2500	−1.2	74.682	14000	−76	13.786
3000	−4.5	70.108	16000	−89	9.632
4000	−11.0	61.640	18000	−102	6.556
5000	−17.5	54.020	2000	−115	4.328

The pressure values in Table 1 may be calculated from

$$p = 101.325(1 - 2.25577 \times 10^{-5} Z)^{5.2559} \tag{3}$$

The equation for temperature as a function of altitude is given as

$$t = 15 - 0.0065 Z \tag{4}$$

where

Z = altitude, m

p = barometric pressure, kPa

part I　Theory

t = temperature, ℃

Equations (3) and (4) are accurate from −5000m to 11000m. For higher altitudes, comprehensive tables of barometric pressure and other physical properties of the standard atmosphere can be found in NASA (1976).

THERMODYNAMIC PROPERTIES OF MOIST AIR

Table 2, developed from formulas by Hyland and Wexler (1983a, b), shows values of thermodynamic properties of **moist air** based on the **thermodynamic temperature scale**. This ideal scale differs slightly from practical temperature scales used for physical measurements. For example, the standard boiling point for water (at 101.325kPa) occurs at 99.97℃ on this scale rather than at the traditional value of 100℃. Most measurements are currently based on the International Temperature Scale of 1990 (ITS-90) (Preston-Thomas 1990).

The following paragraphs briefly describe each column of Table 2:

t = Celsius temperature, based on thermodynamic temperature scale and expressed relative to absolute temperature T in kelvins (K) by the following relation:

$$T = t + 273.15$$

W_s = humidity ratio at saturation, condition at which gaseous phase (moist air) exists in equilibrium with condensed phase (liquid or solid) at given temperature and pressure (standard atmospheric pressure). At given values of temperature and pressure, humidity ratio W can have any value from zero to W_s.

v_{da} = specific volume of dry air, m³/kg (dry air).

$v_{as} = v_s - v_{da}$, difference between specific volume of moist air at saturation and that of dry air itself, m³/kg (dry air), at same pressure and temperature.

v_s = specific volume of moist air at saturation, m³/kg (dry air).

h_{da} = specific enthalpy of dry air, kJ/kg (dry air). In Table 2, h_{da} has been assigned a value of 0 at 0℃ and standard atmospheric pressure.

$h_{as} = h_s - h_{da}$, difference between specific enthalpy of moist air at saturation and that of dry air itself, kJ/kg (dry air), at same pressure and temperature.

h_s = specific enthalpy of moist air at saturation, kJ/kg (dry air).

s_{da} = specific entropy of dry air, kJ/(kg·K) (dry air). In Table 2, s_{da} has been assigned a value of 0 at 0℃ and standard atmospheric pressure.

$s_{as} = s_s - s_{da}$, difference between specific entropy of moist air at saturation and that of dry air itself, kJ/(kg·K) (dry air), at same pressure and temperature.

s_s = specific entropy of moist air at saturation kJ/(kg·K) (dry air).

h_w = specific enthalpy of condensed water (liquid or solid) in equilibrium with saturated moist air at specified temperature and pressure, kJ/kg (water). In Table 2, h_w is assigned a value of 0 at its triple point (0.01℃) and saturation pressure.

Thermodynamic Properties of Moist Air at Standard Atmospheric Pressure, 101.325kPa

Table 2

Temp., ℃ t	Humidity Ratio, kg(w)/kg(da) W_s	Specific Volume, m³/kg(dry air)			Specific Enthalpy, kJ/kg(dry air)			Specific Entropy, kJ/(kg·K)(dry air)			Condensed Water			Temp., ℃ t
											Specific Enthalpy, kJ/kg h_w	Specific Entropy, kJ/(kg·K) s_w	Vapor Pressure, kPa p_s	
		v_{da}	v_{as}	v_s	h_{da}	h_{as}	h_s	s_{da}	s_{as}	s_s				
0*	0.003789	0.7734	0.0047	0.7781	0.000	9.473	9.473	0.0000	0.0364	0.0364	0.06	−0.0001	0.6112	0
1	0.004076	0.7762	0.0051	0.7813	1.006	10.197	11.203	0.0037	0.0391	0.0427	4.28	0.0153	0.6571	1
2	0.004381	0.7791	0.0055	0.7845	2.012	10.970	12.982	0.0073	0.0419	0.0492	8.49	0.0306	0.7060	2
3	0.004707	0.7819	0.0059	0.7878	3.018	11.793	14.811	0.0110	0.0449	0.0559	12.70	0.0459	0.7581	3
4	0.005054	0.7848	0.0064	0.7911	4.024	12.672	16.696	0.0146	0.0480	0.0627	16.91	0.0611	0.8135	4
5	0.005424	0.7876	0.0068	0.7944	5.029	13.610	18.639	0.0182	0.0514	0.0697	21.12	0.0762	0.8725	5
6	0.005818	0.7904	0.0074	0.7978	6.036	14.608	20.644	0.0219	0.0550	0.0769	25.32	0.0913	0.9353	6
7	0.006237	0.7933	0.0079	0.8012	7.041	15.671	22.713	0.0255	0.0588	0.0843	29.52	0.1064	1.0020	7
8	0.006683	0.7961	0.0085	0.8046	8.047	16.805	24.852	0.0290	0.0628	0.0919	33.72	0.1213	1.0729	8
9	0.007157	0.7990	0.0092	0.8081	9.053	18.010	27.064	0.0326	0.0671	0.0997	37.92	0.1362	1.1481	9
10	0.007661	0.8018	0.0098	0.8116	10.059	19.293	29.352	0.0362	0.0717	0.1078	42.11	0.1511	1.2280	10
11	0.008197	0.8046	0.0106	0.8152	11.065	20.658	31.734	0.0397	0.0765	0.1162	46.31	0.1659	1.3128	11
12	0.008766	0.8075	0.0113	0.8188	12.071	22.108	34.179	0.0433	0.0816	0.1248	50.50	0.1806	1.4026	12
13	0.009370	0.8103	0.0122	0.8225	13.077	23.649	36.726	0.0468	0.0870	0.1337	54.69	0.1953	1.4979	13
14	0.010012	0.8132	0.0131	0.8262	14.084	25.286	39.370	0.0503	0.0927	0.1430	58.88	0.2099	1.5987	14
15	0.010692	0.8160	0.0140	0.8300	15.090	27.023	42.113	0.0538	0.0987	0.1525	63.07	0.2244	1.7055	15
16	0.011413	0.8188	0.0150	0.8338	16.096	28.867	44.963	0.0573	0.1051	0.1624	67.26	0.2389	1.8185	16
17	0.012178	0.8217	0.0160	0.8377	17.102	30.824	47.926	0.0607	0.1119	0.1726	71.44	0.2534	1.9380	17
18	0.012989	0.8245	0.0172	0.8417	18.108	32.900	51.008	0.0642	0.1190	0.1832	75.63	0.2678	2.0643	18
19	0.013848	0.8274	0.0184	0.8457	19.114	35.101	54.216	0.0677	0.1266	0.1942	79.81	0.2821	2.1979	19
20	0.014758	0.8302	0.0196	0.8498	20.121	37.434	57.555	0.0711	0.1346	0.2057	84.00	0.2965	2.3389	20
21	0.015721	0.8330	0.0210	0.8540	21.127	39.908	61.035	0.0745	0.1430	0.2175	88.18	0.3107	2.4878	21
22	0.016741	0.8359	0.0224	0.8583	22.133	42.527	64.660	0.0779	0.1519	0.2298	92.36	0.3249	2.6448	22
23	0.017821	0.8387	0.0240	0.8627	23.140	45.301	68.440	0.0813	0.1613	0.2426	96.55	0.3390	2.8105	23
24	0.018963	0.8416	0.0256	0.8671	24.146	48.239	72.385	0.0847	0.1712	0.2559	100.73	0.3531	2.9852	24
25	0.020170	0.8444	0.0273	0.8717	25.153	51.347	76.500	0.0881	0.1817	0.2698	104.91	0.3672	3.1693	25
26	0.021448	0.8472	0.0291	0.8764	26.159	54.638	80.798	0.0915	0.1927	0.2842	109.09	0.3812	3.3633	26
27	0.022798	0.8501	0.0311	0.8811	27.165	58.120	85.285	0.0948	0.2044	0.2992	113.27	0.3951	3.5674	27
28	0.024226	0.8529	0.0331	0.8860	28.172	61.804	89.976	0.0982	0.2166	0.3148	117.45	0.4090	3.7823	28
29	0.025735	0.8558	0.0353	0.8910	29.179	65.699	94.878	0.1015	0.2296	0.3311	121.63	0.4229	4.0084	29

* Extrapolated to represent metastable equilibrium with undercooled liquid.

Note that h_w is greater than the steam-table enthalpy of saturated pure condensed phase by the amount of enthalpy increase governed by the pressure increase from saturation pressure to 101.325kPa, plus influences from presence of air.

s_w = specific entropy of condensed water (liquid or solid) in equilibrium with saturated

part I *Theory*

air, kJ/(kg·K) (water); s_w differs from entropy of pure water at saturation pressure, similar to h_w.

p_s = vapor pressure of water in saturated moist air, kPa. Pressure p_s differs negligibly from saturation vapor pressure of pure water p_{ws} for conditions shown. Consequently, values of p_s can be used at same pressure and temperature in equations where p_{ws} appears. Pressure p_s is defined as $p_s = x_{ws} p$, where x_{ws} is mole fraction of water vapor in moist air saturated with water at temperature t and pressure p, and where p is total barometric pressure of moist air.

THERMODYNAMIC PROPERTIES OF WATER AT SATURATION

Table 3 shows thermodynamic properties of **water at saturation** for temperatures from -60 to 160°C, calculated by the formulations described by Hyland and Wexler (1983b). Symbols in the table follow standard steam table nomenclature. These properties are based on the thermodynamic temperature scale. The enthalpy and entropy of saturated liquid water are both assigned the value zero at the triple point, 0.01°C. Between the triple-point and critical-point temperatures of water, two states—liquid and vapor—may coexist in equilibrium. These states are called **saturated liquid** and **saturated vapor**.

The **water vapor saturation pressure** is required to determine a number of moist air properties, principally the saturation humidity ratio. Values may be obtained from Table 3 or calculated from the following formulas (Hyland and Wexler 1983b).

The saturation pressure over **ice** for the temperature range of -100 to 0°C is given by

$$\ln p_{ws} = C_1/T + C_2 + C_3 T + C_4 T^2 + C_5 T^3 + C_6 T^4 + C_7 \ln T \tag{5}$$

where

$C_1 = -5.6745359\text{E}+03$

$C_2 = 6.3925247\text{E}+00$

$C_3 = -9.6778430\text{E}-03$

$C_4 = 6.2215701\text{E}-07$

$C_5 = 2.0747825\text{E}-09$

$C_6 = -9.4840240\text{E}-13$

$C_7 = -4.1635019\text{E}+00$

The saturation pressure over **liquid water** for the temperature range of 0 to 200°C is given by

$$\ln p_{ws} = C_8/T + C_9 + C_{10} T + C_{11} T^2 + C_{12} T^3 + C_{13} \ln T \tag{6}$$

where

$C_8 = -5.8002206\text{E}+03$

$C_9 = 1.3914993\text{E}+00$

$C_{10} = -4.8640239\text{E}-02$

$C_{11} = 4.1764768\text{E}-05$

$C_{12} = -1.4452093\text{E}-08$

$C_{13} = 6.5459673\text{E}+00$

In both Equations (5) and (6),

ln = natural logarithm

p_{ws} = saturation pressure, Pa

T = absolute temperature, K = ℃ + 273.15

The coefficients of Equations (5) and (6) have been derived from the Hyland-Wexler equations. Due to rounding errors in the derivations and in some computers' calculating precision, the results obtained from Equations (5) and (6) may not agree precisely with Table 3 values.

Thermodynamic Properties of Water at Saturation   Table 3

Temp., ℃ t	Absolute Pressure, kPa p	Specific Volume, m³/kg(water)			Specific Enthalpy, kJ/kg(water)			Specific Entropy, kJ/(kg·K)(water)			Temp., ℃ t
		Sat. Liquid v_f	Evap. v_{fg}	Sat. Vapor v_g	Sat. Liquid h_f	Evap. h_{fg}	Sat. Vapor h_g	Sat. Liquid s_f	Evap. s_{fg}	Sat. Vapor s_g	
0	0.6112	0.001000	206.141	206.143	−0.04	2500.81	2500.77	−0.0002	9.1555	9.1553	0
1	0.6571	0.001000	192.455	192.456	4.18	2498.43	2502.61	0.0153	9.1134	9.1286	1
2	0.7060	0.001000	179.769	179.770	8.39	2496.05	2504.45	0.0306	9.0716	9.1022	2
3	0.7580	0.001000	168.026	168.027	12.60	2493.68	2506.28	0.0459	9.0302	9.0761	3
4	0.8135	0.001000	157.137	157.138	16.81	2491.31	2508.12	0.0611	8.9890	9.0501	4
5	0.8725	0.001000	147.032	147.033	21.02	2488.94	2509.96	0.0763	8.9482	9.0244	5
6	0.9353	0.001000	137.653	137.654	25.22	2486.57	2511.79	0.0913	8.9077	8.9990	6
7	1.0020	0.001000	128.947	128.948	29.42	2484.20	2513.62	0.1064	8.8674	8.9738	7
8	1.0728	0.001000	120.850	120.851	33.62	2481.84	2515.46	0.1213	8.8273	8.9488	8
9	1.1481	0.001000	113.326	113.327	37.82	2479.47	2517.29	0.1362	8.7878	8.9245	9
10	1.2280	0.001000	106.328	106.329	42.01	2477.11	2519.12	0.1511	8.7484	8.8995	10
11	1.3127	0.001000	99.812	99.813	46.21	2474.74	2520.95	0.1659	8.7093	8.8752	11
12	1.4026	0.001001	93.743	93.744	50.40	2472.38	2522.78	0.1806	8.6705	8.8511	12
13	1.4978	0.001001	88.088	88.089	54.59	2470.02	2524.61	0.1953	8.6319	8.8272	13
14	1.5987	0.001001	82.815	82.816	58.78	2467.66	2526.44	0.2099	8.5936	3.8035	14
15	1.7055	0.001001	77.897	77.898	62.97	2465.30	2528.26	0.2244	8.5556	8.7801	15
16	1.8184	0.001001	73.307	73.308	67.16	2462.93	2530.09	0.2389	8.5178	8.7568	16
17	1.9380	0.001001	69.021	69.022	71.34	2460.57	2531.92	0.2534	8.4804	8.7338	17
18	2.0643	0.001002	65.017	65.018	75.53	2458.21	2533.74	0.2678	8.4431	8.7109	18
19	2.1978	0.001002	65.274	61.273	79.72	2455.85	2535.56	0.2821	8.4061	8.6883	19
20	2.3388	0.001002	57.774	57.773	83.90	2453.48	2537.38	0.2964	8.3694	8.6658	20
21	2.4877	0.001002	54.450	54.500	88.08	2451.12	2539.20	0.3107	8.3329	8.6436	21
22	2.6448	0.001002	51.433	51.434	92.27	2448.75	2541.02	0.3249	8.2967	8.6215	22
23	2.8104	0.001003	48.562	48.563	96.45	2446.39	2542.84	0.3390	8.2607	8.5996	23
24	2.9851	0.001003	45.872	45.873	100.63	2444.02	2544.65	0.3531	8.2249	8.5780	24
25	3.1692	0.001003	43.350	43.351	104.81	2441.66	2546.47	0.3672	8.1894	8.5565	25
26	3.3631	0.001003	40.985	40.986	108.99	2439.29	2548.28	0.3812	8.1541	8.5352	26
27	3.5673	0.001004	38.766	38.767	113.18	2436.92	2550.09	0.3951	8.1190	8.5141	27
28	3.7822	0.001004	36.682	36.683	117.36	2434.55	2551.90	0.4090	8.0842	8.4932	28
29	4.0083	0.001004	34.726	34.727	121.54	2432.17	2553.71	0.4229	8.0496	8.4724	29

part Ⅰ　Theory

HUMIDITY PARAMETERS

Basic Parameters

　　Humidity ratio (alternatively, the moisture content or mixing ratio) W of a given moist air sample is defined as the ratio of the mass of water vapor to the mass of dry air contained in the sample:

$$W = M_w / M_{da} \tag{7}$$

The humidity ratio W is equal to the mole fraction ratio x_w/x_{da} multiplied by the ratio of molecular masses, namely, $18.01528/28.9645 = 0.62198$:

$$W = 0.62198 x_w / x_{da} \tag{8}$$

　　Specific humidity γ is the ratio of the mass of water vapor to the total mass of the moist air sample:

$$\gamma = M_w / (M_w + M_{da}) \tag{9a}$$

In terms of the humidity ratio,

$$\gamma = W/(1+W) \tag{9b}$$

　　Absolute humidity (alternatively, water vapor density) d_v is the ratio of the mass of water vapor to the total volume of the sample:

$$d_v = M_w / V \tag{10}$$

The **density** ρ of a moist air mixture is the ratio of the total mass to the total volume:

$$\rho = (M_{da} + M_w)/V = (1/v)(1+W) \tag{11}$$

where v is the moist air specific volume, m³/kg (dry air), as defined by Equation (27).

Humidity Parameters Involving Saturation

　　The following definitions of humidity parameters involve the concept of moist air saturation:

　　Saturation humidity ratio $W_s(t, p)$ is the humidity ratio of moist air saturated with respect to water (or ice) at the same temperature t and pressure p.

　　Degree of saturation μ is the ratio of the air humidity ratio W to the humidity ratio W_s of saturated moist air at the same temperature and pressure:

$$\mu = \left.\frac{W}{W_s}\right|_{t,p} \tag{12}$$

　　Relative humidity ϕ is the ratio of the mole fraction of water vapor x_w in a given moist air sample to the mole fraction x_{ws} in an air sample saturated at the same temperature and pressure:

$$\phi = \left.\frac{x_w}{x_{ws}}\right|_{t,p} \tag{13}$$

Combining Equations (8), (12), and (13),

$$\mu = \frac{\phi}{1+(1-\phi)W_s/0.62198} \tag{14}$$

Dew-point temperature t_d is the temperature of moist air saturated at the same pressure p, with the same humidity ratio W as that of the given sample of moist air. It is defined as the solution $t_d(p, W)$ of the following equation:

$$W_s(p, t_d) = W \tag{15}$$

Thermodynamic wet-bulb temperature t^* is the temperature at which water (liquid or solid), by evaporating into moist air at a given dry-bulb temperature t and humidity ratio W, can bring air to saturation adiabatically at the same temperature t^* while the total pressure p is maintained constant. This parameter is considered separately in the section on Thermodynamic Wet-Bulb Temperature and Dew-Point Temperature.

PERFECT GAS RELATIONSHIPS FOR DRY AND MOIST AIR

When moist air is considered a mixture of independent perfect gases (i.e., dry air and water vapor), each is assumed to obey the perfect gas equation of state as follows:

Dry air: $\quad p_{da}V = n_{da}RT \tag{16}$

Water vapor: $\quad p_w V = n_w RT \tag{17}$

where

p_{da} = partial pressure of dry air

p_w = partial pressure of water vapor

V = total mixture volume

n_{da} = number of moles of dry air

n_w = number of moles of water vapor

R = universal gas constant, 8314.41 J/(kg mol · K)

T = absolute temperature, K

The mixture also obeys the perfect gas equation:

$$pV = nRT \tag{18}$$

or

$$(p_{da} + p_w)V = (n_{da} + n_w)RT \tag{19}$$

where $p = p_{da} + p_w$ is the total mixture pressure and $n = n_{da} + n_w$ is the total number of moles in the mixture. From Equations (16) through (19), the mole fractions of dry air and water vapor are, respectively,

$$x_{da} = p_{da}/(p_{da} + p_w) = p_{da}/p \tag{20}$$

and

$$x_w = p_w/(p_{da} + p_w) = p_w/p \tag{21}$$

From Equations (8), (20), and (21), the **humidity ratio** W is given by

$$W = 0.62198 \frac{p_w}{p - p_w} \tag{22}$$

The degree of saturation μ is, by definition, Equation (12):

$$\mu = \frac{W}{W_s}\bigg|_{t,p}$$

where

$$W_s = 0.62198 \frac{p_{ws}}{p - p_{ws}} \quad (23)$$

The term p_{ws} represents the saturation pressure of water vapor in the absence of air at the given temperature t. This pressure p_{ws} is a function only of temperature and differs slightly from the vapor pressure of water in saturated moist air.

The **relative humidity** ϕ is, by definition, Equation (13):

$$\phi = \frac{x_w}{x_{ws}} \bigg|_{t,p}$$

Substituting Equation (21) for x_w and x_{ws},

$$\phi = \frac{p_w}{p_{ws}} \bigg|_{t,p} \quad (24)$$

Substituting Equation (21) for x_{ws} into Equation (14),

$$\phi = \frac{\mu}{1 - (1-\mu)(p_{ws}/p)} \quad (25)$$

Both ϕ and μ are zero for dry air and unity for saturated moist air. At intermediate states their values differ, substantially so at higher temperatures.

The **specific volume** v of a moist air mixture is expressed in terms of a unit mass of dry air:

$$v = V/M_{da} = V/(28.9645 n_{da}) \quad (26)$$

where V is the total volume of the mixture, M_{da} is the total mass of dry air, and n_{da} is the number of moles of dry air. By Equations (16) and (26), with the relation $p = p_{da} + p_w$,

$$v = \frac{RT}{28.9645(p - p_w)} = \frac{R_{da}T}{p - p_w} \quad (27)$$

Using Equation (22),

$$v = \frac{RT(1 + 1.6078W)}{28.964 p} = \frac{R_{da}T(1 + 1.6078W)}{p} \quad (28)$$

In Equations (27) and (28), v is specific volume, T is absolute temperature, p is total pressure, p_w is the partial pressure of water vapor, and W is the humidity ratio.

In specific units, Equation (28) may be expressed as

$$v = 0.2871(t + 273.15)(1 + 1.6078W)/p$$

where

v = specific volume, m³/kg (dry air)

t = dry-bulb temperature, ℃

W = humidity ratio, kg (water)/kg (dry air)

p = total pressure, kPa

The **enthalpy** of a mixture of perfect gases equals the sum of the individual partial enthalpies of the components. Therefore, the specific enthalpy of moist air can be written as follows:

$$h = h_{da} + W h_g \quad (29)$$

where h_{da} is the specific enthalpy for dry air in kJ/kg (dry air) and h_g is the specific enthalpy for saturated water vapor in kJ/kg (water) at the temperature of the mixture. As an approximation,

$$h_{da} \approx 1.006t \tag{30}$$

$$h_g \approx 2501 + 1.805t \tag{31}$$

where t is the dry-bulb temperature in ℃. The moist air specific enthalpy in kJ/kg (dry air) then becomes

$$h = 1.006t + W(2501 + 1.805t) \tag{32}$$

THERMODYNAMIC WET-BULB TEMPERATURE AND DEW-POINT TEMPERATURE

For any state of moist air, a temperature t^* exists at which liquid (or solid) water evaporates into the air to bring it to saturation at exactly this same temperature and total pressure (Harrison 1965). During the adiabatic saturation process, the saturated air is expelled at a temperature equal to that of the injected water. In this constant pressure process,

- Humidity ratio is increased from a given initial value W to the value W_s^* corresponding to saturation at the temperature t^*;
- Enthalpy is increased from a given initial value h to the value h_s^* corresponding to saturation at the temperature t^*;
- Mass of water added per unit mass of dry air is $(W_s^* - W)$, which adds energy to the moist air of amount $(W_s^* - W)h_w^*$, where h_w^* denotes the specific enthalpy in kJ/kg (water) of the water added at the temperature t^*.

Therefore, if the process is strictly adiabatic, conservation of enthalpy at constant total pressure requires that

$$h + (W_s^* - W)h_w^* = h_s^* \tag{33}$$

The properties W_s^*, h_w^*, and h_s^* are functions only of the temperature t^* for a fixed value of pressure. The value of t^*, which satisfies Equation (33) for given values of h, W, and p, is the **thermodynamic wet-bulb temperature.**

The **psychrometer** consists of two thermometers, one thermometer's bulb is covered by a wick that has been thoroughly wetted with water. When the wet bulb is placed in an airstream, water evaporates from the wick, eventually reaching an equilibrium temperature called the **wet-bulb temperature**. This process is not one of adiabatic saturation, which defines the thermodynamic wet-bulb temperature, but one of simultaneous heat and mass transfer from the wet bulb. The fundamental mechanism of this process is described by the Lewis relation. Fortunately, only small corrections must be applied to wet-bulb thermometer readings to obtain the thermodynamic wet-bulb temperature.

As defined, thermodynamic wet-bulb temperature is a unique property of a given moist air sample independent of measurement techniques.

Equation (33) is exact since it defines the thermodynamic wet-bulb temperature t^*. Substituting the approximate perfect gas relation [Equation (32)] for h, the corresponding expression for h_s^*, and the approximate relation

$$h_w^* \approx 4.186 t^* \tag{34}$$

into Equation (33), and solving for the humidity ratio,

$$W = \frac{(2501 - 2.381 t^*) W_s^* - 1.006(t - t^*)}{2501 + 1.805 t - 4.186 t^*} \tag{35}$$

where t and t^* are in ℃.

The **dew-point temperature** t_d of moist air with humidity ratio W and pressure p was defined earlier as the solution $t_d(p, w)$ of $W_s(p, t_d)$. For perfect gases, this reduces to

$$p_{ws}(t_d) = p_w = (pW)/(0.62198 + W) \tag{36}$$

where p_w is the water vapor partial pressure for the moist air sample and $p_{ws}(t_d)$ is the saturation vapor pressure at temperature t_d. The saturation vapor pressure is derived from Table 3 or from Equation (5) or (6). Alternatively, the dew-point temperature can be calculated directly by one of the following equations (Peppers 1988):

For the dew-point temperature range of 0 to 93℃,

$$t_d = C_{14} + C_{15}\alpha + C_{16}\alpha^2 + C_{17}\alpha^3 + C_{18}(p_w)^{0.1984} \tag{37}$$

For temperatures below 0℃,

$$t_d = 6.09 + 12.608\alpha + 0.4959\alpha^2 \tag{38}$$

where

t_d = dew-point temperature, ℃

$\alpha = \ln p_w$

p_w = water vapor partial pressure, kPa

$C_{14} = 6.54$

$C_{15} = 14.526$

$C_{16} = 0.7389$

$C_{17} = 0.09486$

$C_{18} = 0.4569$

PSYCHROMETRIC CHARTS

A psychrometric chart graphically represents the thermodynamic properties of moist air.

The choice of coordinates for a psychrometric chart is arbitrary. A chart with coordinates of enthalpy and humidity ratio provides convenient graphical solutions of many moist air problems with a minimum of thermodynamic approximations. ASHRAE developed seven such psychrometric charts. Chart No.1 is shown as Figure 1; the others may be obtained through ASHRAE.

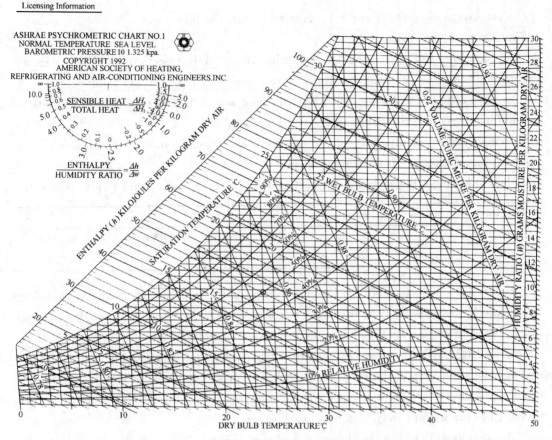

Fig. 1　ASHRAE Psychrometric Chart No. 1

Charts 1 through 4 are for sea level pressure (101.325kPa), Chart 5 is for 750m altitude (92.66kPa), Chart 6 is for 1500m altitude (84.54kPa), and Chart 7 is for 2250m altitude (77.04kPa). All charts use oblique-angle coordinates of enthalpy and humidity ratio, and are consistent with the data of Table 2 and the properties computation methods of Goff and Gratch (1945), and Goff (1949) as well as Hyland and Wexler (1983a). Palmatier (1963) describes the geometry of chart construction applying specifically to Charts 1 and 4.

The dry-bulb temperature ranges covered by the charts are

Charts 1, 5, 6, 7	Normal temperature	0 to 50℃
Chart 2	Low temperature	−40 to 10℃
Chart 3	High temperature	10 to 120℃
Chart 4	Very high temperature	100 to 200℃

Psychrometric properties or charts for other barometric pressures can be derived by interpolation. Sufficiently exact values for most purposes can be derived by methods described in the section on Perfect Gas Relationships for Dry and Moist Air. The construction of charts for altitude conditions has been treated by Haines (1961), Rohsenow

part I Theory

(1946), and Karig (1946).

Comparison of Charts 1 and 6 by overlay reveals the following:
1. The dry-bulb lines coincide.
2. Wet-bulb lines for a given temperature originate at the intersections of the corresponding dry-bulb line and the two saturation curves, and they have the same slope.
3. Humidity ratio and enthalpy for a given dry- and wet-bulb temperature increase with altitude, but there is little change in relative humidity.
4. Volume changes rapidly; for a given dry-bulb and humidity ratio, it is practically inversely proportional to barometric pressure.

The following table compares properties at sea level (Chart 1) and 1500m (Chart 6):

Chart No.	db	wb	h	w	rh	v
1	40	30	99.5	23.0	49	0.920
6	40	30	114.1	28.6	50	1.111

Figure 1, which is ASHRAE Psychrometric Chart No. 1, shows humidity ratio lines (horizontal) for the range from 0 (dry air) to 30g (water)/kg (dry air). Enthalpy lines are oblique lines drawn across the chart precisely parallel to each other.

Dry-bulb temperature lines are drawn straight, not precisely parallel to each other, and inclined slightly from the vertical position. Thermodynamic wet-bulb temperature lines are oblique lines that differ slightly in direction from that of enthalpy lines. They are straight but are not precisely parallel to each other.

Relative humidity lines are shown in intervals of 10%. The saturation curve is the line of 100% rh, while the horizontal line for $W=0$ (dry air) is the line for 0% rh.

Specific volume lines are straight but are not precisely parallel to each other.

A narrow region above the saturation curve has been developed for fog conditions of moist air. This two-phase region represents a mechanical mixture of saturated moist air and liquid water, with the two components in thermal equilibrium. Isothermal lines in the fog region coincide with extensions of thermodynamic wet-bulb temperature lines. If required, the fog region can be further expanded by extension of humidity ratio, enthalpy, and thermodynamic wet-bulb temperature lines.

The protractor to the left of the chart shows two scales—one for sensible-total heat ratio, and one for the ratio of enthalpy difference to humidity ratio difference. The protractor is used to establish the direction of a condition line on the psychrometric chart.

Example 1 illustrates use of the ASHRAE Psychrometric Chart to determine moist air properties.

Example 1. Moist air exists at 40℃ dry-bulb temperature, 20℃ thermodynamic wet-bulb temperature, and 101.325kPa pressure. Determine the humidity ratio, enthalpy, dew-point temperature, relative humidity, and specific volume.

Solution: Locate state point on Chart 1 (Figure 1) at the intersection of 40℃ dry-bulb temperature and 20℃ thermodynamic wet-bulb temperature lines. Read **humidity ratio** $W = 6.5$g (water)/kg (dry air).

The **enthalpy** can be found by using two triangles to draw a line parallel to the nearest enthalpy line [60kJ/kg (dry air)] through the state point to the nearest edge scale. Read $h = 56.7$kJ/kg (dry air).

Dew-point temperature can be read at the intersection of $W = 6.5$g (water)/kg (dry air) with the saturation curve. Thus, $t_d = 7$℃.

Relative humidity ϕ can be estimated directly. Thus, $\phi = 14\%$.

Specific volume can be found by linear interpolation between the volume lines for 0.88 and 0.90m³/kg (dry air). Thus, $v = 0.896$m³/kg (dry air).

WORDS AND EXPRESSIONS

algorithm	n. [数] 运算法则
arbitrary	adj. 任意的，武断的，独裁的，专断的
argon	n. [化] 氩
barometric	adj. 大气压力
coexist	vi. 共存
component	n. 成分；adj. 组成的，构成的
consequently	adv. 从而，因此
contaminant	n. 致污物，污染物
dew	n. 露，露水般的东西，清新
dotted line	点线，虚线
expel	v. 驱逐，开除，排出，发射
helium	n. 氦(化学元素，符号为 He)
humidity ratio	含湿量（湿度比）
intermolecular	adj. [化] 分子间的，存在(或作用)于分子间的
krypton	n. [化] 氪
methane	n. [化] 甲烷，沼气
miscellaneous	adj. 各色各样混在一起，混杂的，多才多艺的
neon	n. [化] 氖
nomenclature	n. 命名法，术语
oblique	adj. 倾斜的，间接的，不坦率的，无诚意的
overlay	n. 覆盖，覆盖图
parameter	n. 参数，参量，〈口〉起限定作用的因素
perfect gas	理想气体
pollen	n. 花粉；vt. 传授花粉给
precision	n. 精确，精密度，精度
protractor	n. 量角器

part I Theory

psychrometric chart	n. 焓湿图
psychrometry	焓湿学
relative humidity	相对湿度
stratosphere	n. [气] 同温层，最上层，最高阶段
sulfur	n. [化] 硫磺，硫黄；vt. 用硫磺处理
superintendent	n. 主管，负责人，指挥者，管理者
troposphere	n. [气] 对流层
ultrafine	adj. 极其细小的，非常细微的
xenon	n. 氙（惰性气体的一种，元素符号 Xe）

NOTATIONS

(1) Some Definitions of Humidity：

1) **Humidity ratio** (alternatively, the moisture content or mixing ratio) W of a given moist air sample is defined as the ratio of the mass of water vapor to the mass of dry air contained in the sample.

一给定湿空气样品的湿度比（或者含湿量、混合比）W 定义为包含在本样品中水蒸气质量与干空气质量之比。

2) **Specific humidity** γ is the ratio of the mass of water vapor to the total mass of the moist air sample.

比湿度 γ 是水蒸气质量与整个湿空气样品质量之比。

3) **Absolute humidity** (alternatively, water vapor density) dv is the ratio of the mass of water vapor to the total volume of the sample.

绝对湿度（或者水蒸气密度）dv 是水蒸气质量与总的样品体积之比。

4) **Relative humidity** is the ratio of the mole fraction of water vapor in a given moist air sample to the mole fraction in an air sample saturated at the same temperature and pressure.

相对湿度是一给定湿空气样品中水蒸气的摩尔分数与相同温度和压力下饱和空气样品中水蒸气的摩尔分数之比。

(2) t_d and t^*：

1) **Dew-point temperature** is the temperature of moist air saturated at the same pressure, with the same humidity ratio as that of the given sample of moist air.

露点温度是在相同压强下饱和的湿空气温度，与给定湿空气样品具有相同的湿度比。

2) **Thermodynamic wet-bulb temperature** t^* is the temperature at which water (liquid or solid), by evaporating into moist air at a given dry-bulb temperature and humidity ratio, can bring air to saturation adiabatically at th same temperature t^* while the total pressure p is maintained constant.

热力学湿球温度 t^* 是指水（液态或固态）在该温度下，通过蒸发进入具有给定的干球温度和湿度比的湿空气，能在总的压强不变时使空气在同样的温度 t^* 下由绝热变为饱和。

(3) Harrison (1965) lists the approximate percentage composition of dry air by volume as: nitrogen, 78.084; oxygen, 20.9476; argon, 0.934; carbon dioxide, 0.0314; neon, 0.001818; helium, 0.000524; methane, 0.00015; sulfur dioxide, 0 to 0.0001; hydrogen, 0.00005; and minor components such as krypton, xenon, and ozone, 0.0002.

Harrison (1965) 列出了干空气中各个组分的近似体积百分比：氮，78.084；氧，20.9476；氩，0.934；二氧化碳，0.0314；氖，0.001818；氦，0.000524；甲烷，0.00015；二氧化硫，0 到 0.0001；氢，0.00005；以及一些微小组分如氪、氙和臭氧等 0.0002。

(Pay attention to the pronunciation of the components of dry air.)

(4) Unless otherwise stated, saturation refers to a flat interface surface between the moist air and the condensed phase.

除非另有说明，饱和指的是湿空气和凝相间的平界面。

(5) The temperature is assumed to decrease linearly with increasing altitude throughout the troposphere (lower atmosphere), and to be constant in the lower reaches of the stratosphere.

假设在整个对流层中（较低的大气层），温度随海拔的增加而线性降低，在同温层的较低区域则是常数。

(6) Between the triple-point and critical-point temperatures of water, two states—liquid and vapor—may coexist in equilibrium. These states are called saturated liquid and saturated vapor.

在水的三相点和临界点温度之间，两种状态——液态和气态——可能平衡共存。这些状态称为饱和液态和饱和蒸汽。

(7) The saturation pressure over ice for the temperature range of −100 to 0℃ is given by Equation (5).

在零下 100℃ 到 0℃ 的温度范围内，冰上面的饱和压力由方程(5)给出。

(8) Psychrometric properties or charts for other barometric pressures can be derived by interpolation. Sufficiently exact values for most purposes can be derived by methods described in the section on Perfect Gas Relationships for Dry and Moist Air.

其他大气压下的焓湿性质或线图可以通过插值得到。大多数场合，足够精确的值可以通过干湿空气的理想气体方程所述的方法得到。

(9) The protractor to the left of the chart shows two scales—one for sensible-total heat ratio, and one for the ratio of enthalpy difference to humidity ratio difference.

线图左边的分度尺示有两个刻度，一个是显热——总热比，另一个是焓差和湿度比差之比。

EXERCISES

(1) 请将下列句子翻译成英文。

1) 饱和湿度比 $W_s(t, p)$ 是饱和湿空气的湿度比和相同温度 t、相同压力 p 下水（或冰）

之比。

2）饱和度 μ 是空气湿度比 W 和相同温度压力下的饱和湿空气湿度比 W_s 的比值。

3）湿度比从某一给定的初值 W 增大到温度 t^* 下的饱和状态的值 W_s^*。

4）焓从一给定初值 h 增大到温度 t^* 下饱和态的值 h_s^*。

5）单位质量干空气加入水的质量为 $(W_s^* - W)$，其加给湿空气带进的能量为 $(W_s^* - W)h_w^*$，h_w^* 定义为加入温度 t^* 的水的比焓，单位是 kJ/kg（水）。

(2) 请将下列句子翻译成中文。

1) Psychrometrics deals with thermodynamic properties of moist air and uses these properties to analyze conditions and processes involving moist air, Hyland and Wexler (1983a, 1983b) developed formulas for thermodynamic properties of moist air and water. Perfect gas relations can be used in most air-conditioning problems instead of these formulas. Threlkeld (1970) showed that errors are less than 0.7% in calculating humidity ratio, enthalpy, and specific volume of saturated air at standard atmospheric pressure for a temperature range of −50℃. Furthermore, these errors decrease with decreasing pressure.

2) The psychrometer consists of two thermometers, one thermometer's bulb is covered by a wick that has been thoroughly wetted with water. When the wet bulb is placed in an airstream, water evaporates from the wick, eventually reaching an equilibrium temperature called the wet-bulb temperature. This process is not one of adiabatic saturation, which defines the thermodynamic wet-bulb temperature, but is one of simultaneous heat and mass transfer from the wet bulb.

3) Wet-bulb lines for a given temperature originate at the intersections of the corresponding dry-bulb line and the two saturation curves, and they have the same slope.

Part II

General Engineering Information

LESSON 6

THERMAL COMFORT

A PRINCIPAL purpose of heating, ventilating, and air-conditioning systems is to provide conditions for human thermal comfort. A widely accepted definition is, "Thermal Comfort is that condition of mind that expresses satisfaction with the thermal environment" (ASHRAE *Standard* 55). This definition leaves open what is meant by condition of mind or satisfaction, but it correctly emphasizes that the judgment of comfort is a cognitive process involving many inputs influenced by physical, physiological, psychological, and other processes.

The conscious mind appears to reach conclusions about thermal comfort and discomfort from direct temperature and moisture sensations from the skin, deep body temperatures, and the efforts necessary to regulate body temperatures (Hensel 1973, 1981; Hardy et al. 1971; Gagge 1937; Berglund 1995). In general, comfort occurs when body temperatures are held within narrow ranges, skin moisture is low, and the physiological effort of regulation is minimized.

Comfort also depends on behavioral actions that are initiated unconsciously or by the conscious mind and guided by thermal and moisture sensations to reduce discomfort. Some of the possible behavioral actions to reduce discomfort are altering clothing, altering activity, changing posture or location, changing the thermostat setting, opening a window, complaining, or leaving the space.

Surprisingly, although regional climate conditions, living conditions, and cultures differ widely throughout the world, the temperature that people choose for comfort under like conditions of clothing, activity, humidity, and air movement has been found to be very similar (Fanger 1972; de Dear et al. 1991; Busch 1992).

This Lesson summarizes the fundamentals of human thermoregulation and comfort in terms useful to the engineer for operating systems and designing for the comfort and health of building occupants.

HUMAN THERMOREGULATION

The metabolic activities of the body result almost completely in heat that must be continuously dissipated and regulated to maintain normal body temperatures. Insufficient heat

---

Extracted from Chapter 8 of the ASHRAE *Handbook-Fundamentals*.

loss leads to overheating, also called **hyperthermia**, and excessive heat loss results in body cooling, also called **hypothermia**. Skin temperature greater than 45℃ or less than 18℃ causes pain (Hardy et al. 1952). Skin temperatures associated with comfort at sedentary activities are 33 to 34℃ and decrease with increasing activity (Fanger 1968). In contrast, internal temperatures rise with activity. The temperature regulatory center in the brain is about 36.8℃ at rest in comfort and increases to about 37.4℃ when walking and 37.9℃ when jogging. An internal temperature less than about 28℃ can lead to serious cardiac arrhythmia and death, and a temperature greater than 46℃ can cause irreversible brain damage. Therefore, the careful regulation of body temperature is critical to comfort and health.

The heat produced by a resting adult is about 100W. Because most of this heat is transferred to the environment through the skin, it is often convenient to characterize metabolic activity in terms of heat production per unit area of skin. For the resting person, this is about $58W/m^2$ and is called 1 met. This is based on the average male European, with a skin surface area of about $1.8m^2$. For comparison, female Europeans have an average surface area of $1.6m^2$. Systematic differences in this parameter may occur between ethnic and geographical groups. Higher metabolic rates are often described in terms of the resting rate. Thus, a person working at metabolic rate five times the resting rate would have a metabolic rate of 5 met.

The **hypothalamus**, located in the brain, is the central control organ for body temperature. It has hot and cold temperature sensors and is bathed by arterial blood. Since the recirculation rate of blood in the body is rapid and returning blood is mixed together in the heart before returning to the body, arterial blood is indicative of the average internal body temperature. The hypothalamus also receives thermal information from temperature sensors in the skin and perhaps other locations as well (spinal cord, gut), as summarized by Hensel (1981).

The hypothalamus controls various physiological processes of the body to regulate body temperature. Its control behavior is primarily proportional to deviations from setpoint temperatures with some integral and derivative response aspects. The most important and often used of the physiological processes is regulating blood flow to the skin. When internal temperatures rise above a set point, an increasing proportion of the total blood is directed to the skin. This **vasodilation** of skin blood vessels can increase skin blood flow by 15 times (from $1.7mL/(s \cdot m^2)$ at resting comfort to $25mL/(s \cdot m^2)$ in extreme heat) to carry internal heat to the skin for transfer to the environment. When body temperatures fall below the set point, skin blood flow is reduced (vasoconstricted) to conserve body heat. The effect of maximum vasoconstriction is equivalent to the insulating effect of a heavy sweater. At temperatures less than the set point, muscle tension increases to generate additional heat; where muscle groups are opposed, this may increase to visible shivering. Shivering can double the resting rate of heat production.

At elevated internal temperatures, sweating occurs. This defense mechanism is a powerful way to cool the skin and increase heat loss from the core. The sweating function of the skin and its control is more advanced in humans than in other animals and is increasingly necessary for comfort at metabolic rates above resting level (Fanger 1968). Sweat glands pump perspiration onto the skin surface for evaporation. If conditions are good for evaporation, the skin can remain relatively dry even at high sweat rates with little perception of sweating. At skin conditions less favorable for evaporation, the sweat must spread out on the skin about the sweat gland until the sweat-covered area is sufficient to evaporate the sweat coming to the surface. The fraction of the skin that is covered with water to account for the observed total evaporation rate is termed **skin wettedness**(Gagge 1937).

Humans are quite good at sensing skin moisture from perspiration (Berglund and Cunningham 1986; Berglund 1994), and skin moisture correlates well with warm discomfort and unpleasantness (Winslow et al. 1937). It is rare for a sedentary or slightly active person to be comfortable with a skin wettedness greater than 25%. In addition to the perception of skin moisture, skin wettedness increases the friction between skin and fabrics, making clothing feel less pleasant and fabrics feel more coarse (Gwosdow et al. 1987). This also occurs with architectural materials and surfaces, particularly smooth, nonhygroscopic surfaces.

With repeated intermittent heat exposure, the set point for the onset of sweating decreases and the proportional gain or temperature sensitivity of the sweating system increases (Gonzalez et al. 1978, Hensel 1981). However, under long-term exposure to hot conditions, the set point increases, perhaps to reduce the physiological effort of sweating. Perspiration as secreted has a lower salt concentration than interstitial body fluid or blood plasma. After prolonged heat exposure, sweat glands further reduce the salt concentration of sweat to conserve salt.

At the surface, the water in sweat evaporates while the dissolved salt and other constituents remain and accumulate. Because salt lowers the vapor pressure of water and thereby impedes its evaporation, the accumulating salt results in increased skin wettedness with time. Some of the relief and pleasure of washing after a warm day is related to the restoration of a hypotonic sweat film and decreased **skin wettedness.** Other adaptations to heat are increased blood flow and sweating in peripheral regions where heat transfer is better. Such adaptations are examples of **integral control.**

The role of thermoregulatory effort in comfort is highlighted by the experiments of Chatonnet and Cabanac (1965) and observations of Kuno (1995). Chatonnet's experiments compared the sensation of placing the subject's hand in relatively hot or cold water (30 to 38℃) for 30s given the subject at different thermal states. When the person was overheated or hyperthermic, the cold water was pleasant and the hot water was very unpleasant, but when the subject was in a cold or hypothermic state, the hand felt pleasant in hot water and unpleasant in cold water. Kuno (1995) describes similar observations during transient whole body exposures to hot and cold environment. When a subject is in a state of thermal discomfort, any move away from the thermal stress of the uncomfortable environment is perceived as pleasant during the transition.

ENERGY BALANCE

Figure 1 shows the thermal interaction of the human body with its environment. The total metabolic rate of work M produced within the body is the metabolic rate required for the person's activity M_{act} plus the metabolic level required for shivering M_{shiv} (should shivering occur). A portion of the body's energy production may be expended as external work done by the muscles W; the net heat production $M-W$ is either stored (S), causing the body's temperature to rise, or dissipated to the environment through the skin surface (q_{sk}) and respiratory tract (q_{res}).

$$M-W = q_{sk} + q_{res} + S$$
$$= (C+R+E_{sk}) + (C_{res}+E_{res}) + (S_{sk}+S_{cr}) \tag{1}$$

where

M = rate of metabolic heat production, W/m²

W = rate of mechanical work accomplished, W/m²

q_{sk} = total rate of heat loss from skin, W/m²

q_{res} = total rate of heat loss through respiration, W/m²

$C+R$ = sensible heat loss from skin, W/m²

E_{sk} = total rate of evaporative heat loss from skin, W/m²

C_{res} = rate of convective heat loss from respiration, W/m²

E_{res} = rate of evaporative heat loss from respiration, W/m²

S_{sk} = rate of heat storage in skin compartment, W/m²

S_{cr} = rate of heat storage in core compartment, W/m²

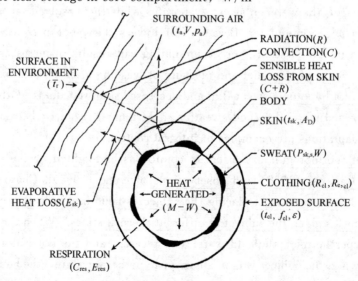

Fig. 1 Thermal Interaction of Human Body and Environment

Heat dissipation from the body to the immediate surroundings occurs by several modes of heat exchange: sensible heat flow $C+R$ from the skin; latent heat flow from the evaporation of sweat E_{rsw} and from evaporation of moisture diffused through the skin E_{dif}; sensible heat flow during respiration C_{res}; and latent heat flow due to evaporation of moisture during respiration E_{res}. Sensible heat flow from the skin may be a complex mixture of conduction, convection, and radiation for a clothed person; however, it is equal to the sum of the convection C and radiation R heat transfer at the outer clothing surface (or exposed skin).

Sensible and latent heat losses from the skin are typically expressed in terms of environmental factors, skin temperature t_{sk}, and skin wettedness w. The expressions also incorporate factors that account for the thermal insulation and moisture permeability of clothing. The independent environmental variables can be summarized as air temperature t_a, mean radiant temperature $\bar{t}_r$, relative air velocity V, and ambient water vapor pressure p_a. The independent personal variables that influence thermal comfort are activity and clothing.

The rate of heat storage in the body equals the rate of increase in internal energy. The body can be considered as two thermal compartments, the skin and the core (see the section on Two-Node Model under Prediction of Thermal Comfort). The rate of storage can be written separately for each compartment in terms of thermal capacity and time rate of change of temperature in each compartment:

$$S_{cr} = \frac{(1-\alpha_{sk})mc_{p,b}}{A_D} \frac{dt_{cr}}{d\theta} \qquad (2)$$

$$S_{sk} = \frac{\alpha_{sk}mc_{p,b}}{A_D} \frac{dt_{sk}}{d\theta} \qquad (3)$$

where

α_{sk} = fraction of body mass concentrated in skin compartment
m = body mass, kg
$c_{p,b}$ = specific heat capacity of body = 3490 J/(kg·K)
A_D = DuBois surface area, m²
t_{cr} = temperature of core compartment, ℃
t_{sk} = temperature of skin compartment, ℃
θ = time, s

The fractional skin mass α_{sk} depends on the rate $\dot{m}_{bl}$ of blood flowing to the skin surface.

CONDITIONS FOR THERMAL COMFORT

In addition to the previously discussed independent environmental and personal variables influencing thermal response and comfort, other factors may also have some effect. These factors, such as nonuniformity of the environment, visual stimuli, age, and outdoor climate are generally considered secondary factors. Studies by Rohles and Nevins (1971) and Rohles (1973) on 1600 college-age students revealed correlations between comfort level, temperature, humidity, sex, and length of exposure. Many of these correlations are given in Table 1. The thermal sensation scale developed for these studies is called the **ASHRAE thermal sensation scale**:

+3	hot
+2	warm
+1	slightly warm
0	neutral
−1	slightly cool
−2	cool
−3	cold

The equations in Table 1 indicate that the women of this study were more sensitive to temperature and less sensitive to humidity than the men. But in general about a 3 K change in temperature or a 3 kPa change in water vapor pressure is necessary to change a thermal sensation vote by one unit or temperature category.

Equations for Predicting Thermal Sensation (Y) of Men, Women, and Men and Women Combined   Table 1

Exposure Period, h	Subjects	Regression Equations[a,b] t = dry-bulb temperature, ℃ p = vapor pressure, kPa
1.0	Men	$Y = 0.220t + 0.233p - 5.673$
	Women	$Y = 0.272t + 0.248p - 7.245$
	Both	$Y = 0.245t + 0.248p - 6.475$
2.0	Men	$Y = 0.221t + 0.270p - 6.024$
	Women	$Y = 0.283t + 0.210p - 7.694$
	Both	$Y = 0.252t + 0.240p - 6.859$
3.0	Men	$Y = 0.212t + 0.293p - 5.949$
	Women	$Y = 0.275t + 0.255p - 8.622$
	Both	$Y = 0.243t + 0.278p - 6.802$

[a] Y values refer to the ASHRAE thermal sensation scale.

[b] For young adult subjects with sedentary activity and wearing clothing with a thermal resistance of approximately 0.5 clo, $\bar{t}_r \approx \bar{t}_a$ and air velocities < 0.2 m/s.

Current and past studies are periodically reviewed to update ASHRAE *Standard* 55, Thermal Environmental Conditions for Human Occupancy. This standard specifies conditions or comfort zones where 80% of sedentary or slightly active persons find the environment thermally acceptable.

Because people typically change their clothing for the seasonal weather, ASHRAE

Standard 55 specifies summer and winter comfort zones appropriate for clothing insulation levels of 0.5 and 0.9 clo (0.078 and 0.14m² · K/W), respectively (Figure 2) (Addendum 55a to ASHRAE *Standard* 55). The warmer and cooler temperature borders of the comfort zones are affected by humidity and coincide with lines of constant ET*. In the middle region of a zone, a typical person wearing the prescribed clothing would have a thermal sensation at or very near neutral. Near the boundary of the warmer zone, a person would feel about +0.5 warmer on the ASHRAE thermal sensation scale; near the boundary of the cooler zone, that person may have a thermal sensation of −0.5.

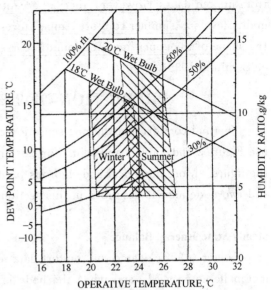

Fig. 2 ASHRAE Summer and Winter Comfort Zones
(Acceptable ranges of operative temperature and humidity for people in typical summer and winter clothing during primarily sedentary activity.)

Comfort zones for other clothing levels can be approximated by decreasing the temperature borders of the zone by 0.6 K for each 0.1 clo increase in clothing insulation and vice versa. Similarly a zone's temperatures can be decreased by 1.4 K per met increase in activity above 1.2 met.

The upper and lower humidity levels of the comfort zones are less precise. Low humidity can lead to drying of the skin and mucous surfaces. Comfort complaints about dry nose, throat, eyes, and skin occur in low-humidity conditions, typically when the dew point is less than 0℃. Liviana et al. (1988) found eye discomfort increased with time in low-humidity environments (dew point<2℃). Green (1982) quantified that respiratory illness and absenteeism increase in winter with decreasing humidity and found that any increase in humidity from very low levels decreased absenteeism in winter. In compliance with these and other discomfort observations, ASHRAE *Standard* 55 recommends that the dew-point temperature of occupied spaces not be less than 2℃.

At high humidity levels, too much skin moisture tends to increase discomfort (Gagge 1937, Berglund and Cunningham 1986), particularly skin moisture that is physiological in origin (water diffusion and perspiration). At high humidity levels, thermal sensation alone is not a reliable predictor of thermal comfort (Tanabe et al. 1987). The discomfort appears to be due to the feeling of the moisture itself, increased friction between skin and clothing with skin moisture (Gwosdow et al. 1986), and other factors. To prevent warm discomfort, Nevins et al. (1975) recommended that on the warm side of the comfort zone the relative humidity not exceed 60%.

The upper humidity limits of ASHRAE *Standard* 55 were developed theoretically from limited data. However, thermal acceptability data gathered at medium and high humidity levels at summer comfort temperatures with subjects wearing 0.55 clo corroborated the shape of the upper limit and found it corresponded to an 80% thermal acceptability level(Berglund 1995).

PREDICTION OF THERMAL COMFORT

Thermal comfort and thermal sensation can be predicted several ways. One way is to use Figure 2 and Table 1 and adjust for clothing and activity levels that differ from those of the figure. More numerical and rigorous predictions are possible by using the PMV-PPD and two-node models described in this section.

Steady-State Energy Balance

Fanger (1982) related the comfort data to physiological variables. At a given level of metabolic activity M, and when the body is not far from thermal neutrality, the mean skin temperature t_{sk} and sweat rate E_{rsw} are the only physiological parameters influencing the heat balance. However, heat balance alone is not sufficient to establish thermal comfort. In the wide range of environmental conditions where heat balance can be obtained, only a narrow range provides thermal comfort. The following linear regression equations based on data from Rohles and Nevins (1971) indicate values of t_{sk} and E_{rsw} that provide thermal comfort.

$$t_{sk,req} = 35.7 - 0.0275(M-W) \tag{4}$$

$$E_{rsw,req} = 0.42(M-W-58.15) \tag{5}$$

At higher activity levels, sweat loss increases and the mean skin temperature decreases. Both reactions increase the heat loss from the body core to the environment. These two empirical relationships link the physiological and heat flow equations and thermal comfort perceptions.

Fanger (1982) reduced these relationships to a single equation, which assumed all sweat generated is evaporated, eliminating clothing permeation efficiency i_{cl} as a factor in the equation. This assumption is valid for normal indoor clothing worn in typical indoor environments with low or moderate activity levels. At higher activity levels ($M_{act} > 3$met), where a significant amount of sweating occurs even at optimum comfort conditions, this assumption may limit accuracy. The reduced equation is slightly different from the heat transfer equations developed here. The radiant heat exchange is expressed in terms of the Stefan-Boltzmann law (instead of using h_r), and diffusion of water vapor through the skin is expressed as a diffusivity coefficient and a linear approximation for saturated vapor pressure evaluated at t_{sk}. The combination of environmental and personal variables that produces a neutral sensation may be expressed as follows:

$$M-W=3.96\times10^{-8}f_{cl}[(t_{cl}+273)^4-(\bar{t}_r+273)^4]+f_{cl}h_c(t_{cl}-t_a)+3.05[5.73-0.007$$
$$(M-W)-p_a]+0.42[(M-W)-58.15]+0.0173M(5.87-p_a)+0.0014M$$
$$(34-t_a) \qquad (6)$$

where

$$t_{cl}=35.7-0.0275(M-W)-R_{cl}\{(M-W)-3.05[5.73-0.007(M-W)-p_a]-0.42$$
$$[(M-W)-58.15]-0.00173M(5.87-p_a)-0.0014M(34-t_a)\} \qquad (7)$$

The values of h_c and f_{cl} can be estimated from tables and equations given in the section on Engineering Data and Measurements. Fanger used the following relationships:

$$h_c=\begin{cases} 2.38(t_{cl}-t_a)^{0.25} & 2.38(t_{cl}-t_a)^{0.25}>12.1\sqrt{V} \\ 12.1\sqrt{V} & 2.38(t_{cl}-t_a)^{0.25}<12.1\sqrt{V} \end{cases} \qquad (8)$$

$$f_{cl}=\begin{cases} 1.0+0.2I_{cl} & I_{cl}<0.5\text{clo} \\ 1.05+0.1I_{cl} & I_{cl}>0.5\text{clo} \end{cases} \qquad (9)$$

Figures 3 and 4 show examples of how Equation (6) can be used.

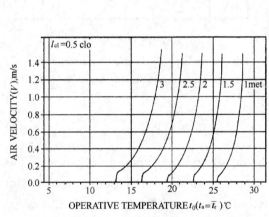

Fig. 3 Air Velocities and Operative Temperatures at 50% rh Necessary for Comfort (PMV=0) of Persons in Summer Clothing at Various Levels of Activity

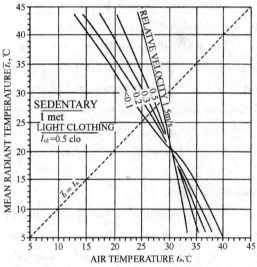

Fig. 4 Air Temperatures and Mean Radiant Temperatures Necessary for Comfort (PMV=0) of Sedentary Persons in Summer Clothing at 50% rh

Equation (6) is expanded to include a range of thermal sensations by using a **predicted mean vote (PMV) index**. The PMV index predicts the mean response of a large group of people according to the ASHRAE thermal sensation scale. Fanger (1970) related PMV to the imbalance between the actual heat flow from the body in a given environment and the heat flow required for optimum comfort at the specified activity by the following equation:

$$\text{PMV}=[0.303\exp(-0.036M)+0.028]L \qquad (10)$$

where L is the thermal load on the body, defined as the difference between internal heat production and heat loss to the actual environment for a person hypothetically kept at comfort values of t_{sk} and E_{rsw} at the actual activity level. Thermal load L is then the difference between the left and right sides of Equation (6) calculated for the actual values of the environmental conditions. As part of this calculation, the clothing temperature t_{cl} is found by iteration as

$$t_{cl} = 35.7 - 0.028(M-W) - R_{cl}\{39.6 \times 10^{-9} f_{cl}[(t_{cl}+273)^4 - (\bar{t}_r+273)^4] + f_{cl}h_c(t_{cl}-t_a)\} \quad (11)$$

After estimating the PMV with Equation (10) or another method, the **predicted percent dissatisfied** (PPD) with a condition can also be estimated. Fanger (1982) related the PPD to the PMV as follows:

$$PPD = 100 - 95\exp[-(0.03353PMV^4 + 0.2179PMV^2)] \quad (12)$$

where dissatisfied is defined as anybody not voting -1, $+1$, or 0. This relationship is shown in Figure 13. A PPD of 10% corresponds to the PMV range of ± 0.5, and even with PMV=0, about 5% of the people are dissatisfied.

The **PMV-PPD model** is widely used and accepted for design and field assessment of comfort conditions. ISO *Standard* 7730 includes a short computer listing that facilitates computing PMV and PPD for a wide range of parameters.

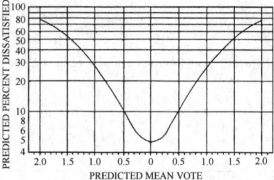

Fig. 5 Predicted Percentage of Dissatisfied (PPD) as Function of Predicted Mean Vote (PMV)

Two-Node Model

The PMV model is useful only for predicting steady-state comfort responses. The two-node model can be used to predict physiological responses or responses to transient situations, at least for low and moderate activity levels in cool to very hot environments (Gagge et al. 1971, 1986). The two-node model is a simplification of more complex thermoregulatory models developed by Stolwijk and Hardy (1966). The simple, lumped parameter model considers a human as two concentric thermal compartments that represent the skin and the core of the body.

The **skin compartment** simulates the epidermis and dermis and is about 1.6mm thick. Its mass, which is about 10% of the total body, depends on the amount of blood flowing through it for thermoregulation. The temperature in a compartment is assumed to be uniform so that the only temperature gradients are between compartments. In a cold environment, blood flow to the extremities may be reduced to conserve the heat of vital organs, resulting in axial temperature gradients in the arms, legs, hands, and feet. Heavy exercise with certain muscle groups or asymmetric environmental conditions may also cause nonuniform compartment temperatures and limit the accuracy of the model.

All the heat is assumed to be generated in the **core compartment.** In the cold, shivering and muscle tension may generate additional metabolic heat. This increase is related to skin and core temperature depressions from their set point values, or

$$M_{shiv}=19.4(34-t_{sk})(37-t_{cr}) \tag{13}$$

where the temperature terms are set to zero if they become negative.

The core loses energy when the muscles do work on the surroundings. Heat is also lost from the core through respiration. The rate of respiratory heat loss is due to sensible and latent changes in the respired air and the ventilation rate.

In addition, heat is conducted passively from the core to the skin. This is modeled as a massless thermal conductor $[K=5.28\text{W}/(\text{m}^2 \cdot \text{K})]$. A controllable heat loss path from the core consists of pumping variable amounts of warm blood to the skin for cooling. This peripheral blood flow Q_{bl} in $\text{L/h} \cdot \text{m}^2$ depends on skin and core temperature deviations from their respective set points:

$$Q_{bl}=\frac{\text{BFN}+c_{dil}(t_{cr}-37)}{1+S_{tr}(34-t_{sk})} \tag{14}$$

The temperature terms can only be >0. If the deviation is negative, the term is set to zero. For average persons, the coefficients BFN, c_{dil}, and S_{tr} are 6.3, 175 and 0.5. Further, skin blood flow Q_{bl} is limited to a maximum of 90 $\text{L}/(\text{h} \cdot \text{m}^2)$.

Dry (sensible) heat loss q_{dry} from the skin flows through the clothing by conduction and then by parallel paths to the air and surrounding surfaces. Evaporative heat follows a similar path, flowing through the clothing and through the air boundary layer. Maximum evaporation E_{max} occurs if the skin is completely covered with sweat. The actual evaporation rate E_{sw} depends on the size w of the sweat film:

$$E_{sw}=wE_{max} \tag{15}$$

where w is E_{rsw}/E_{max}.

The rate of regulatory sweating E_{rsw} (rate at which water is brought to the surface of the skin in W/m^2) can be predicted by skin and core temperature deviations from their set points:

$$E_{rsw}=c_{sw}(t_b-t_{bset})\exp[-(t_{sk}-34)/10.7] \tag{16}$$

where $t_b=(1-\alpha_{sk})t_{cr}+\alpha_{sk}t_{sk}$ and is the mean body temperature, and $c_{sw}=170\text{W}/(\text{m}^2 \cdot \text{K})$. The temperature deviation terms are set to zero when negative. α_{sk} is the fraction of the total body mass that is considered to be thermally in the skin compartment.

$$\alpha_{sk}=0.0418+\frac{0.745}{10.8Q_{bl}-0.585} \tag{17}$$

Regulatory sweating Q_{rsw} in the model is limited to $1\text{L}/(\text{h} \cdot \text{m}^2)$ or 670 W/m^2. E_{rsw} evaporates from the skin, but if E_{rsw} is greater than E_{max}, the excess drips off.

An energy balance on the core yields

$$M + M_{shiv} = W + q_{res} + (K + SKBF c_{p,bl})(t_{cr} - t_{sk}) + m_{cr} c_{cr} \frac{dt_{cr}}{d\theta} \tag{18}$$

and for the skin,

$$(K + SKBF c_{p,bl})(t_{cr} - t_{sk}) = q_{dry} + q_{evqp} + m_{sk} c_{sk} \frac{dt_{sk}}{d\theta} \tag{19}$$

where c_{cr}, c_{sk}, and $c_{p,bl}$ are specific heats of core, skin, and blood [3500, 3500, and 4190 J/(kg·K), respectively], and SKBF is $\rho_{bl} Q_{bl}$, where ρ_{bl} is density of blood (12.9 kg/L).

Equations (18) and (19) can be rearranged in terms of $dt_{sk}/d\theta$ and $dt_{cr}/d\theta$ and numerically integrated with small time steps (10 to 60s) from initial conditions or previous values to find t_{cr} and t_{sk} at any time.

After calculating values of t_{sk}, t_{cr}, and w, the model uses empirical expressions to predict thermal sensation (TSENS) and thermal discomfort (DISC). These indices are based on 11-point numerical scales, where positive values represent the warm side of neutral sensation or comfort, and negative values represent the cool side. TSENS is based on the same scale as PMV, but with extra terms for ±4 (very hot/cold) and ±5 (intolerably hot/cold). Recognizing the same positive/negative convention for warm/cold discomfort, DISC is defined as

5   intolerable
4   limited tolerance
3   very uncomfortable
2   uncomfortable and unpleasant
1   slightly uncomfortable but acceptable
0   comfortable

TSENS is defined in terms of deviations of mean body temperature t_b from cold and hot set points representing the lower and upper limits for the zone of evaporative regulation: $t_{b,c}$ and $t_{b,h}$, respectively. The values of these set points depend on the net rate of internal heat production and are calculated by

$$t_{b,c} = \frac{0.194}{58.15}(M - W) + 36.301 \tag{20}$$

$$t_{b,h} = \frac{0.347}{58.15}(M - W) + 36.669 \tag{21}$$

TSENS is then determined by

$$\text{TSENS} = \begin{cases} 0.4685(t_b - t_{b,c}) & t_b < t_{b,c} \\ 4.7\eta_{ev}(t_b - t_{b,c})/(t_{b,h} - t_{b,c}) & t_{b,c} \leqslant t_b \leqslant t_{b,h} \\ 4.7\eta_{ev} + 0.4685(t_b - t_{b,h}) & t_{b,h} < t_b \end{cases} \tag{22}$$

where η_{ev} is the evaporative efficiency (assumed to be 0.85).

DISC is numerically equal to TSENS when t_b is below its cold set point $t_{b,c}$ and it is related to skin wettedness when body temperature is regulated by sweating:

$$\text{DISC} = \begin{cases} 0.4685(t_b - t_{p,c}) & t_b < t_{b,c} \\ \dfrac{4.7(E_{rsw} - E_{rsw,req})}{E_{max} - E_{rsw,req} - E_{dif}} & t_{b,c} \leq t_b \end{cases} \tag{23}$$

where $E_{rsw,req}$ is calculated as in Fanger's model, using Equation(5).

Adaptive Models

Adaptive models do not actually predict comfort responses but rather the almost constant conditions under which people are likely to be comfortable in buildings. In general, people naturally adapt and may also make various adjustments to themselves and their surroundings to reduce discomfort and physiological strain. It has been observed that, through adaptive actions, an acceptable degree of comfort in residences and offices is possible over a range of air temperatures from about 17 to 31℃ (Humphreys and Nicol 1998).

The adaptive adjustments are typically conscious behavioral actions such as altering clothing, posture, activity schedules, activity levels, rate of working, diet, ventilation, air movement, and local temperature. The adaptations may also include unconscious longer term changes to physiological set points and gains for the control of shivering, skin blood flow, and sweating, as well as adjustments to body fluid levels and salt loss. However, only limited documentation and information on such changes is available.

An important driving force behind the adaptive process is the pattern of outside weather conditions and the exposure to them. This is the principal input to the adaptive models that have evolved to date, and these models predict likely comfort temperatures t_c or ranges of t_c from monthly mean outdoor temperatures t_{out}. Such a model (Humphreys and Nicol 1998), based on data from a wide range of buildings, climates, and cultures is

$$t_c = 24.2 + 0.43(t_{out} - 22)\exp-\left(\dfrac{t_{out} - 22)}{24\sqrt{2}}\right)^2 \tag{24}$$

The adaptive models are useful to guide design and energy decisions. They may also be useful to specify building temperatures set points throughout the year. A recent ASHRAE-sponsored study on adaptive models compiled an extensive database from past field studies to study, develop, and test adaptive models. For climates and buildings where cooling and central heating are not required, the study suggests the following model (de Dear and Brager 1998):

$$t_{0c} = 18.9 + 0.255 t_{out} \tag{25}$$

where t_{0c} is the operative comfort temperature.

In general, the value of using an adaptive model to specify set points or guide temperature control strategies is likely to increase with the freedom that occupants are given to adapt (e.g., by having flexible working hours, locations, or dress codes).

Zones of Comfort and Discomfort

The section on Two-Node Model shows that comfort and thermal sensation are not necessarily the same variable, especially for a person in the zone of evaporative thermal regulation. Figures 6 and 7 show this difference for the standard combination of metclo-air movement used in the standard effective temperature. Figure 6 demonstrates that practically all basic physiological variables predicted by the two-node model are functions of ambient temperature and are relatively independent of vapor pressure. All exceptions occur at relative humidities above 80% and as the isotherms reach the ET* =41.5℃ line, where regulation by evaporation fails. Figure 7 shows that lines of constant ET* and wettedness are functions of both ambient temperature and vapor pressure. Thus, human thermal responses are divided into two classes—those in Figure 6, which respond only to heat stress from the environment, and those in Figure 7, which respond to both the heat stress from the environment and the resultant heat strain(Stolwijk et al. 1968).

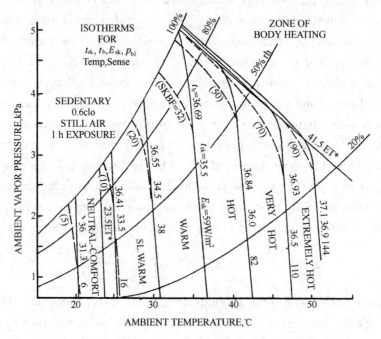

Fig. 6 Effect of Environmental Conditions on Physiological Variables

For warm environments, any index with isotherms parallel to skin temperature is a reliable index of thermal sensation alone, and not of discomfort caused by increased humidity. Indices with isotherms parallel to ET* are reliable indicators of discomfort or dissatisfaction with thermal environments. For a fixed exposure time to cold, lines of constant t_{sk}, ET*, and t_0 are essentially identical, and cold sensation is no different from cold discomfort. For a state of comfort with sedentary or light activity, lines of constant t_{sk} and ET* coincide. Thus comfort and thermal sensations coincide in this region as well. The upper and lower temperature limits for comfort at these levels can be specified either by thermal sensation (Fanger 1982) or by ET*, as is done in ASHRAE *Standard* 55,

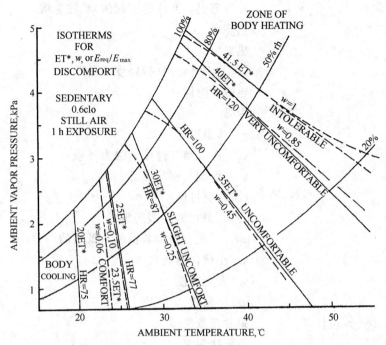

Fig. 7　Effect of Thermal Environment on Discomfort

since lines of constant comfort and lines of constant thermal sensation should be identical.

WORDS AND EXPRESSIONS

absenteeism	n. 缺勤
absorptance	n. 吸收系数
adipose	adj. 脂肪的，肥胖的
arrhythmia	n. [医] 心律不齐，心律失常
cardiovascular	adj. 心脏血管的
constituent	n. 成分，组分
crouch	vi. 蜷缩，蹲伏
curvature	n. 弯曲，曲率
decrement	n. 消耗
derivative	adj. 导数的；n. 导数
dermis	n. 真皮，皮肤
DuBois surface area	人体杜波依斯外表面积
effective radiant field	有效辐射场
endocrine	n. 内分泌
epidermis	n. [生] 表皮，上皮
extremities	n. 四肢，手足
frostbite	n. 霜寒，冻伤

gooseflesh	n. （恐惧，寒冷等引起的）鸡皮疙瘩
heat cramp	热痉挛
heat debt	热损失
heat exhaustion	［医］中暑衰竭（指轻度中暑）
heat stress index	热应力指标
heat stroke	中暑
hyperthermia	n. 体温过高
hyperventilation	n. ［医］换气过度，强力呼吸
hypothalamus	n. ［解剖］视丘下部
hypothermia	n. 体温过低
hypothetically	adv. 假设地，假想地
hypotonic	adj. ［医］张力减退的，低渗的
indisposition	n. 小病，不适宜
irradiance	n. 发光密度
isotherm	n. 等温线
lumped parameter model	集总参数模型
metabolic rate	新陈代谢率
meteorologist	n. 气象学者
molar	adj. 摩尔的
mucous	adj. 黏液的，黏液似的
National Institute of Occupational Safety and Health (NIOSH)	美国国家人员安全与健康研究院
nonhygroscopic	adj. 不吸湿性的，不收湿的
norepinephrine	n. 降肾上腺素，去甲肾上腺素
operative temperature	折算温度，操作温度
perception	n. 感知，感觉
peripheral	adj. 外围的
permeability	n. 渗透性
perspiration	n. 排汗
pigment	vt. 染色
plasma	n. ［解］血浆
postmortem	adv. 死后的，死后发生的
predicted mean vote (PMV)	预测平均投票
predicted percent dissatisfied (PPD)	预测不满意百分数
projection area	投射面积
proportional gain	比例增益
psychological	adj. 心理的，心理学的
psychrometric	adj. 焓湿的
radiometric	adj. 放射分析的

receptor	*n.*	接受器，感受器，受体
rectal	*adj.*	直肠的
resting	*adj.*	静止的
roughness	*n.*	粗糙度
ruff	*n.*	环状领
sauna	*n.*	桑拿浴，蒸汽浴
secrete	*vt.*	分泌
spinal cord	*n.*	脊髓
splanchnic	*adj.*	内脏的
stimuli	*n.*	刺激物
sweat gland	*n.*	汗腺
synergistic	*adj.*	协同作用的
thermal inertia		热惯性
thermal sensation scale		热感觉等级
thermoregulation	*n.*	［生理］体温调节，温度调节
torso		人体躯干
vasoconstriction	*n.*	［生理］血管收缩
vasodilation	*n.*	［生理］血管舒张
vasomotor	*adj.*	血管收缩的

NOTATIONS

(1) The hypothalamus controls various physiological processes of the body to regulate body temperature. Its control behavior is primarily proportional to deviations from set-point temperatures with some integral and derivative response aspects.

下丘脑通过控制人体的各种生理活动来调节体温。其控制行为主要以某种积分和微分的响应特征与设定温度的偏离成比例。

(2) The fraction of the skin that is covered with water to account for the observed total evaporation rate is termed **skin wettedness**.

以被汗水覆盖的皮肤面积百分率来计算观测到的总蒸发率称为皮肤湿度。

(3) The warmer and cooler temperature borders of the comfort zones are affected by humidity and coincide with lines of constant ET*.

热舒适区域的温度（热或冷）边界受湿度的影响，并且与等有效温度 ET 线相重合。

(4) Thus, human thermal responses are divided into two classes—those in Figure 6, which respond only to heat stress from the environment, and those in Figure 7, which respond to both the heat stress from the environment and the resultant heat strain.

因此，人体热反应可以分为两类：一类如图 6 所示，仅对环境的热应力有反应；一类如图 7 所示，对环境的热应力和相应的热应变均有反应。

part II　General Engineering Information

(5) 注意下列句子中的动词及其短语的用法：

1) This Lesson summarizes(概述，总结)the fundamentals of human thermoregulation and comfort in terms useful to the engineer for operating systems and designing for the comfort and health of building occupants.

2) This is based on(基于)the average male European, with a skin surface area of about 1.8m^2.

3) Higher metabolic rates are often described(被描述为) in terms of the resting rate.

4) The fraction of the skin that is covered with water to account for the observed total evaporation rate is termed(被称为)skin wettedness.

5) This also occurs(发生)with architectural materials and surfaces, particularly smooth, nonhygroscopic surfaces.

6) Some of the relief and pleasure of washing after a warm day is related to(与…有关系) the restoration of a hypotonic sweat film and decreased skin wettedness.

7) The role of thermoregulatory effort in comfort is highlighted(使…显得重要)by the experiments of Chatonnet and Cabanac (1965) and observations of Kuno (1995).

8) Figure 1 shows(给出，示出) the thermal interaction of the human body with its environment.

9) Sensible and latent heat losses from the skin are typically expressed(被表示为)in terms of environmental factors, skin temperature t_{sk}, and skin wettedness w.

10) The rate of heat storage in the body equals(等于)the rate of increase in internal energy.

11) The rate of storage can be written(被写为)separately for each compartment in terms of thermal capacity and time rate of change of temperature in each compartment.

EXERCISES

(1) 请指出下列句子中"subject"所表示的含义。

1) Chatonnet's experiments compared the sensation of placing the subject's hand in relatively hot or cold water (30 to 38℃) for 30 s given the subject at different thermal states.

2) However, thermal acceptability data gathered at medium and high humidity levels at summer comfort temperatures with subjects wearing 0.55 clo corroborated the shape of the upper limit and found it corresponded to an 80% thermal acceptability level.

(2) 请用英语表达。

干球温度，湿球温度，黑球温度，露点温度

LESSON 7

AIR CONTAMINANTS

AIR is composed mainly of gases. The major gaseous components of clean, dry air near sea level are approximately 21% oxygen, 78% nitrogen, 1% argon, and 0.04% carbon dioxide.

Normal outdoor air contains varying amounts of foreign materials (permanent atmospheric impurities). These materials can arise from natural processes such as wind erosion, sea spray evaporation, volcanic eruption, and metabolism or decay of organic matter. The natural contaminant concentrations in the air that we breathe vary but are usually lower than those caused by human activity.

Man-made outdoor contaminants are many and varied, originating from numerous types of human activity. Electric power-generating plants, various modes of transportation, industrial processes, mining and smelting, construction, and agriculture generate large amounts of contaminants.

Contaminants that present particular problems in the indoor environment include, among others, tobacco smoke, radon, and formaldehyde.

Air composition may be changed accidentally or deliberately. In sewers, sewage treatment plants, tunnels, and mines, the oxygen content of air can become so low that people cannot remain conscious or survive. Concentrations of people in confined spaces (theaters, survival shelters, submarines) require that carbon dioxide given off by normal respiratory functions be removed and replaced with oxygen. Pilots of high-altitude aircraft, breathing at greatly reduced pressure, require systems that increase oxygen concentration. Conversely, for divers working at extreme depths, it is common to increase the percentage of **helium** in the atmosphere and reduce nitrogen and sometimes oxygen concentrations.

At atmospheric pressure, oxygen concentrations less than 12% or carbon dioxide concentrations greater than 5% are dangerous, even for short periods. Lesser deviations from normal composition can be hazardous under prolonged exposures.

CLASSES OF AIR CONTAMINANTS

The major classes of air contaminants are particulate and gaseous. The **particulate**

Extracted from Chapter 12 of the ASHRAE *Handbook-Fundamentals*.

class covers a vast range of particle sizes from dust large enough to be visible to the eye to submicroscopic particles that elude most filters. Particulates may be solid or liquid. The following traditional contaminant classifications are subclasses of particulates:

- **Dusts**, **fumes**, and **smokes**, which are mostly solid particulate matter, although smoke often contains liquid particles
- **Mists**, **fogs**, and **smogs**, which are mostly suspended liquid particles smaller than those in dusts, fumes, and smokes
- **Bioaerosols**, including viruses, bacteria, fungal spores, and pollen, whose primary impact is related to their biological origin
- Particle size definitions such as **coarse** or **fine**, **visible** or **invisible**, and **macroscopic**, **microscopic**, or **submicroscopic**
- Definitions that relate to particle interaction with the human respiratory system, such as **inhalable** and **respirable**

These classes, their characteristics, units of measurement, and measurement methods are discussed in more detail in this chapter.

The **gaseous** class covers chemical contaminants that can exist as free molecules or atoms in air. Molecules and atoms are smaller than particles and may behave differently as a result. This class covers two important subclasses:

- **Gases**, which are naturally gaseous under ambient indoor or outdoor conditions
- **Vapors**, which are normally solid or liquid under ambient indoor or outdoor conditions, but which evaporate readily

Through evaporation, liquids change into vapors and mix with the surrounding atmosphere. Like gases, they are formless fluids that expand to occupy the space or enclosure in which they are confined. Typical gaseous contaminants, their characteristics, units of measurement, and measurement methods are discussed in detail later in this chapter.

Air contaminants can also be classified according to their sources; their properties; or the health, safety, and engineering issues faced by people exposed to them. Any of these can form a convenient classification system because they allow grouping of applicable standards, guidelines, and control strategies. Most of the following classes include both particulate and gaseous contaminants. The classes are

- Industrial air contaminants
- Nonindustrial indoor air contaminants (including indoor air quality)
- Flammable gases and vapors
- Combustible dusts
- Radioactive contaminants
- Soil gases

LESSON 7

PARTICULATE CONTAMINANTS

PARTICULATE MATTER

Airborne particulate matter is not a single substance but a complex mixture of many different components, usually from several different sources. Particles can be anthropogenic or natural in origin. **Anthropogenic** particles are those produced by human activities, including fossil fuel combustion, industrial processes, and road dust. Particles from **natural** sources do not involve human activity. They include wind-blown dust and smoke from forest fires.

Particles can be generated by primary or secondary processes. Particles from **primary processes** are those emitted directly into the air. Particles from **secondary processes** are those formed from condensable vapors and chemical reactions.

Details on the different particle types are given below.

Types of Solid Particles

Dusts are solid particles projected into the air by natural forces such as wind, volcanic eruption, or earthquakes, or by mechanical processes including crushing, grinding, demolition, blasting, drilling, shoveling, screening, and sweeping. Some of these forces produce dusts by reducing larger masses, while others disperse materials that have already been reduced. Particles are not considered to be dust unless they are smaller than about $100\mu m$. Dusts can be mineral, such as rock, metal, or clay; vegetable, such as grain, flour, wood, cotton, or pollen; or animal, including wool, hair, silk, feathers, and leather.

Fumes are solid particles formed by condensation of vapors of solid materials. Metallic fumes are generated from molten metals and usually occur as oxides because of the highly reactive nature of finely divided matter. Fumes can also be formed by sublimation, distillation, or chemical reaction. Such processes create airborne particles smaller than $1\mu m$. Fumes permitted to age may agglomerate into larger clusters.

Bioaerosols include airborne viruses, bacteria, pollen, and fungus spores. **Viruses** range in size from 0.003 to $0.06\mu m$, although they usually occur in colonies or attached to other particles. Most **bacteria** range between 0.4 and $5\mu m$ and are usually associated with large particles. **Fungus** spores are usually 2 to $10\mu m$, while **pollen** grains are 10 to $100\mu m$, with many common varieties in the 20 to $40\mu m$ range.

Types of Liquid Particles

Mists are small airborne droplets of materials that are ordinarily liquid at normal temperatures and pressure. They can be formed by atomizing, spraying, mixing, violent chemical reactions, evolution of gas from liquid, or escape as a dissolved gas when pressure is released. Small droplets expelled or atomized by sneezing constitute mists.

Fogs are fine airborne droplets, usually formed by condensation of vapor, which remain airborne longer than mists. Fog nozzles are named for their ability to produce extra fine droplets, as compared with mists from ordinary spray devices. Many droplets in fogs or clouds are microscopic and submicroscopic and serve as a transition stage between larger mists and vapors.

The volatile nature of most liquids reduces the size of their airborne droplets from the mist to the fog range and eventually to the vapor phase, until the air becomes saturated with that liquid. If solid material is suspended or dissolved in the liquid droplet, it remains in the air as particulate contamination. For example, sea spray evaporates fairly rapidly, generating a large number of fine salt particles that remain suspended in the atmosphere.

Complex Particles

Smokes are small solid and/or liquid particles produced by incomplete combustion of organic substances such as tobacco, wood, coal, oil, and other carbonaceous materials. The term smoke is applied to a mixture of solid, liquid, and gaseous products, although technical literature distinguishes between such components as soot or carbon particles, fly ash, cinders, tarry matter, unburned gases, and gaseous combustion products. Smoke particles vary in size, the smallest being much less than $1 \mu m$ in diameter. The average is often in the range of 0.1 to $0.3 \mu m$.

Environmental tobacco smoke consists of a suspension of 0.01 to $1.0 \mu m$ (mass median diameter of $0.3 \mu m$) liquid particles that form as the superheated vapors leaving the burning tobacco condense. Also produced are numerous gaseous contaminants including carbon monoxide.

Smog commonly refers to air pollution; it implies an air mixture of smoke particles, mists, and fog droplets of such concentration and composition as to impair visibility, in addition to being irritating or harmful. The composition varies among different locations and at different times. The term is often applied to haze caused by a sunlight-induced photochemical reaction involving the materials in automobile exhausts. <u>Smog is often associated with temperature inversions in the atmosphere that prevent normal dispersion of contaminants.</u>

Sizes of Airborne Particles

Particle size can be defined in several different ways. These depend, for example, on the source or method of generation, the visibility, or the effects.

Particles can be classified as coarse or fine. Coarse particles are larger, and are generally formed by mechanical breaking up of solids. They generally have a minimum size of 1 to $3 \mu m$ (EPA 1996). Fine particles are generally formed from chemical reactions or condensing gases. These particles have a maximum size of about 1 to 3μm. Fine particles are usually more chemically complex, anthropogenic, and secondary in origin, while the coarse particles are predominantly primary, natural, and chemically inert. Coarse particles also include bioaerosols such as mold spores, pollen, animal dander, and dust mite particles that can affect the immune system.

The differences between fine and coarse particles lead to a bimodal distribution of particles in most environments with concentration peaks at about 0.25μm and 5μm. Figure 1 shows a typical urban distribution including the chemical species present in each mode.

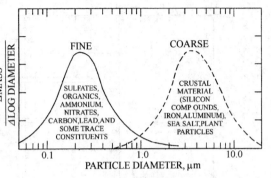

Fig. 1 Typical Urban Aerosol Composition by Particle Size Fraction
(EPA 1982, Willeke and Baron 1993)

The size of a particle determines where in the human respiratory system the particle is deposited. The **inhalable mass** is made up of particles that may deposit anywhere in the respiratory system and is represented by a sample with a median cut point of 100μm. The **thoracic particle mass** is that fraction which can penetrate to the lung airways and is represented by a sample with a median cut point of 10μm. The **respirable particulate mass** is that fraction that can penetrate to the gas-exchange region of the lungs and is represented by a sample with a median cut point of 4μm (ACGIH 1998). Figure 2 illustrates the relative deposition efficiencies of various sizes of particles. The characteristics described above were used by the EPA in development of their outdoor PM_{10} and $PM_{2.5}$ standards.

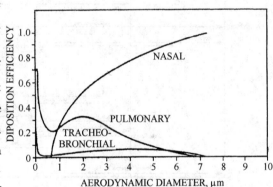

Fig. 2 Relative Deposition Efficiencies of Different Sized Particles in the Three Main Regions of the Human Respiratory System, Calculated for Moderate Activity Level
(Task Group on Lung Dynamics 1966)

Most particles are irregular in shape, and it is useful to characterize their size in terms of some standard particle. For example, the **aerodynamic (equivalent) diameter** of a particle is defined as the diameter of a unit-density sphere having the same gravitational settling velocity as the particle in question (Willeke and Baron 1993).

The tendency of particles to settle on surfaces is a property of interest. Figure 3 shows the sizes of typical indoor airborne solid and liquid particles. Particles smaller than 0.1μm behave like gas molecules exhibiting Brownian motion due to collisions with air molecules and having no measurable settling velocity. Particles in the range from 0.1 to

$1\mu m$ have settling velocities that can be calculated but are so low that settling is usually negligible, since normal air currents counteract any settling. By number, over 99.9% of the particles in a typical atmosphere are below $1\mu m$ (this means fewer than 1 particle in every 1000 is larger than $1\mu m$.).

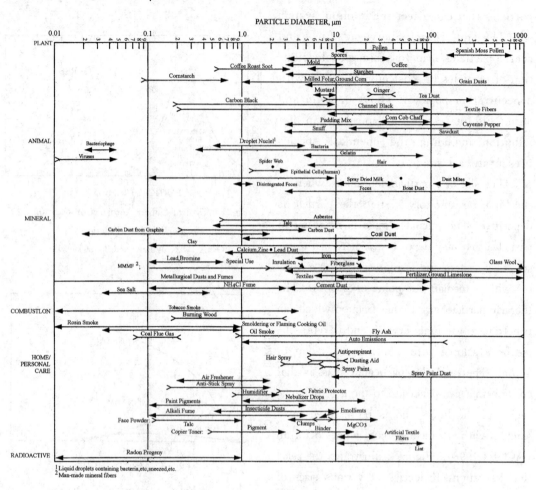

Fig. 3 Sizes of Indoor Particles
(Owen et al. 1992)

Particles in the 1 to $10\mu m$ range settle in still air at constant and appreciable velocity. However, normal air currents keep them in suspension for appreciable periods.

Particles larger than $10\mu m$ settle fairly rapidly and can be found suspended in air only near their source or under strong wind conditions. Exceptions are lint and other light fibrous materials, such as portions of certain weed seeds, which remain suspended longer. Table 1 shows settling times for various types of particles.

Most individual particles $10\mu m$ or larger are visible to the naked eye under favorable conditions of lighting and contrast. Smaller particles are visible only in high concentrations. Cigarette smoke (with an average particle size less than $0.5\mu m$) and clouds are common examples.

Approximate Particle Sizes and Their Times to Settle One Metre — Table 1

Type of Particle	Diameter, μm	Settling Time
Human hair	100 to 150	5s
Skin flakes	20 to 40	
Observable dust in air	>10	
Common pollens	15 to 25	
Mite allergens	10 to 20	5min
Common spores	2 to 10	
Bacteria	1 to 5	
Cat dander	1±0.5	10h
Tobacco smoke	0.1 to 1	
Metal and organic fumes	<0.1 to 1	
Cell debris	0.01 to 1	
Viruses	<0.1	10days

Source: J. D. Spengler, Harvard School of Public Health.

Direct fallout in the vicinity of the dispersing stack or flue and other nuisance problems of air pollution involve larger particles. Smaller particles, as well as mists, fogs, and fumes, remain in suspension longer. In this size range, meteorology and topography are more important than physical characteristics of the particles. Since settling velocities are small, the ability of the atmosphere to disperse these small particles depends largely on local weather conditions.

Comparison is often made to screen sizes used for grading useful industrial dusts and granular materials. Table 2 illustrates the relationship of U.S. standard sieve mesh to particle size in micrometres. Particles above $40\mu m$ are the screen sizes, and those below are the subscreen or microscopic sizes.

Relation of Screen Mesh to Particle Size — Table 2

U. S. Standard sieve mesh	400	325	200	140	100	60	35	18
Nominal sieve opening, μm	37	44	74	105	149	250	500	1000

Particle Size Distribution

The particle size distribution in any sample can be expressed as the percentage of the number of particles smaller than a specified size, area, or mass. The upper curve of Figure 4 shows these data plotted for typical atmospheric contamination. The middle curve shows the percentage of the total projected area of the particles contributed by particles less than a specified size. The lower curve shows the percentage of the total particle mass contributed by thos eparticles less than a given size. Thus, for particles $10\mu m$ and larger, 4% by mass, a 90% arrestance filter essentially removes 90% of 4% for an efficiency by mass of 3.6%.

part II General Engineering Information

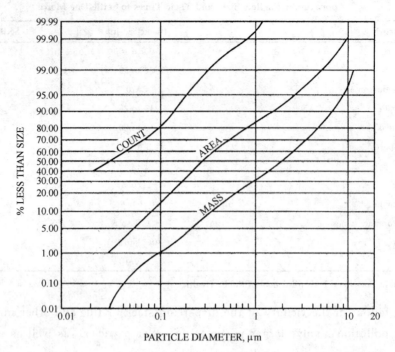

Count curve: Based on measurements by electron microscope
Area curve: Calculated
Mass curve: Solid section based on measurements by Sedimentation
Fig. 4 Particle Size Distribution of Atmospheric Dust
(Whitby et al. 1955, 1957)

The differences among values presented by the three curves should be noted. For example, particles 0.1μm or less in diameter (but still above electron microscope minimum detection size of about 0.005μm) make up 80% of the number of particles in the atmosphere but contribute only 1% of the mass. Also, the 0.1% of particles by number larger than 1μm carry 70% of the total mass, which is the direct result of the mass of a spherical particle increasing as the cube of its diameter. Although most of the mass is contributed by intermediate and larger particles, over 80% of the area (staining) contamination is supplied by particles less than 1μm in diameter, which is in the center of the respirable particle size range and is the size most likely to remain in the lungs. Of possible concern to the HVAC industry is the fact that most of the staining effect on ceilings, walls, windows, and light fixtures, as well as fouling of heat transfer devices and rotating equipment, results from particles less than 1μm in diameter. Suspended particles in urban air are predominantly smaller than 1μm (aerodynamic diameter) and have a distribution that is approximately log-normal.

Units of Measurement

The quantity of particulate matter in the air can be determined as mass or particle count in a given volume of air. Mass units are milligrams per cubic metre of air sampled (mg/m^3) or micrograms per cubic metre of air sampled ($\mu g/m^3$). $1mg/m^3 = 1000\mu g/m^3$. Particle counts are usually quoted for volumes of $0.1ft^3$, $1ft^3$, or 1 litre and are specified for a given range of particle diameter.

Measurement of Airborne Particles

Suitable methods for determining the quantity of particulate matter in the air vary depending on the amount present and on the size of particles involved.

The particle mass (total or respirable) is easily determined by measuring the mass of a filter before and after drawing a known volume of dusty air through it. This method is widely used in industrial workplaces, where there may be significant numbers of large particles, but is not sensitive enough for evaluating office environments.

Optical particle counters are widely used and likely to become more so since ASHRAE adopted *Standard* 52.2. This standard defines a laboratory method for assessing the performance of media filters using an optical particle counter to measure particle counts upstream and downstream of the filter in 12 size ranges between 0.3 and $10\mu m$. Filters are then given a **minimum efficiency reporting value**(MERV) rating based on the count data.

Counters are also used to test cleanrooms for compliance with Federal *Standard* 209E and ISO *Standard* 14644-1. Cleanrooms are defined in terms of the number of particles in certain size ranges that they contain.

Optical particle counters use laser light-scattering to continuously count and size airborne particles and can detect particles down to $0.1\mu m$(ASTM *Standard* F 50). A **condensation nucleus counter** can count particles to below $0.01\mu m$. These particles, present in great numbers in the atmosphere, serve as nuclei for condensation of water vapor (Scala 1963).

Another indirect method measures the **optical density** of the collected dust based on the projected area of the particles. Dust particles can be sized with graduated scales or optical comparisons using a standard **microscope.** The lower limit for sizing with the light field microscope is approximately $0.9\mu m$, depending on the vision of the observer, the dust color, and the contrast available. This size can be reduced to about $0.4\mu m$ by using oil immersion objective techniques. Dark field microscopic techniques reveal particles smaller than these, to a limit of approximately $0.1\mu m$. Smaller submicroscopic dusts can be sized and compared with the aid of an electron microscope. Other sizing techniques may take into account velocity of samplings in calibrated devices and actual settlement measurements in laboratory equipment. The electron microscope and various sampling instruments such as the **cascade impactor** have been successful in sizing particulates, including fogs and mists.

Each of the various methods of measuring particle size distribution gives a different value for the same size particle, since different properties are actually measured. For example, a microscopic technique may measure longest dimension, while impactor results are based on aerodynamic behavior(ACGIH 2001).

Typical Particle Levels

Particle counters, which detect particles larger than about $0.1\mu m$, have indicated that the number of suspended particles is enormous. A room with heavy cigarette smoke has a particle concentration of 10^9 particles per cubic metre. Even clean air typically contains over 35×10^6 particles/m^3. If smaller particles detectable by other means, such as an electron microscope or condensation nucleus counter, are also included, the total particle concentration would be greater than the above concentrations by a factor of 10 to 100.

Extensive measurements have been made of outdoor pollution, but limited data have been gathered on indoor pollution not associated with specific industrial processes. Indoor levels are influenced by the number of people and their activities, building materials and construction, outside conditions, ventilation rate, and the air-conditioning and filtration system. For further information, see the section on Nonindustrial Indoor Air Contaminants, Spengler et al. (1982), and NRC (1981).

BIOAEROSOLS

Bioaerosols are airborne microbiological particulate matter derived from fungi, bacteria, viruses, protozoa, algae, pollen, mites, and their cellular or cell mass components. Bioaerosols are universally present in both indoor and outdoor environments. Problems of concern to engineers occur when microorganisms grow and reproduce indoors.

Microorganisms break down complex molecules found in dead organic materials to simple substances such as carbon dioxide, water, and nitrates. These components are then used by photosyn-thetic organisms such as plants and algae. Thus, the presence of bacteria and fungi in soil, water, and atmospheric habitats is normal. Spores of *Cladosporium*, a fungus commonly found on leaves and dead vegetation, are almost always found in outdoor air samples. They are found in variable numbers in indoor air, depending on the amount of outdoor air that infiltrates into interior spaces or is brought in by the HVAC system. Outdoor microorganisms can also enter on shoes and clothing and be transferred to other surfaces in buildings.

Public interest has focused on airborne microorganisms responsible for diseases and infections, primarily bacteria and viruses. A variety of airborne microorganisms are of economic significance and can cause product contamination or loss. In the food processing industry, yeast and mold can reduce the shelf life of some products. Refined syrups can be damaged by mold scums. Wild yeast can destroy a batch of beer. Antibiotic yields can be reduced by foreign organisms in the culture mix.

Fungi

Much attention has been given to fungi, which include yeasts, molds, and mildews, as well as large mushrooms, puffballs, and bracket fungi. All fungi depend on external sources of organic material for both energy requirements and carbon skeletons. Thus, they cannot increase in number unless supplied with a suitable food source such as small quantities of dust, paper, or wood. Reproduction also requires appropriate temperatures (20 to 40℃) and the presence of water, high air humidity(typically greater than 60%), and/or high moisture content in the food source.

Bacteria

Cooling towers, evaporative condensers, and domestic water service systems all provide water and nutrients for amplification of microorganisms such as *Legionella pneumophila*. Growth of microbial populations to excessive concentrations is generally associated with inadequate preventive maintenance of these systems, at least in cooling towers. A body of literature has identified characteristics of indoor plumbing and heating systems associated with frequent isolation of *Legionella* species, including blind ends, scale, upright electric water heaters, and lower water temperatures. The survival of *Legionella* is enhanced by a variety of parameters including but not limited to warm temperatures, particular algal and protozoan associations, and symbiotic relationships with certain aquatic plants(Fliermans 1985). Evidence has indicated that amoebae and other protozoa act as natural hosts and amplifiers for *Legionella* in the environment (Barbaree et al. 1986). ASHRAE (1989) has further information on the topic.

Bacteria from the soil are likely to be spore-formers and are capable of surviving in hostile environments. Other airborne bacteria, especially within closed occupied spaces, may originate from droplet nuclei caused by actions such as sneezing or be carried on human or animal skin scales.

Pollen

Pollen grains discharged by weeds, grasses, and trees (Hewson et al. 1967, Jacobson and Morris 1977, Solomon and Mathews 1978) and capable of causing hay fever have properties of special interest to air-cleaning equipment designers. Whole grains and fragments transported by air range between 10 and 50μm; however, some measure as small as 5μm, and others measure over 100μm in diameter. Ragweed pollen grains are fairly uniform in size, ranging from 15 to 25μm.

Most pollen grains are hygroscopic and, therefore, vary in mass with the humidity. Illustrations and data on pollen grains are available in botanical literature. Geographical distribution of plants that produce hay fever is also recorded.

The quantity of pollen in the air is generally estimated by exposing an adhesive-coated glass plate outdoors for 24 h, then counting calibrated areas under the microscope. Methods are available for determining the number of pollen grains in a measured volume of air. However, despite their greater accuracy, these methods have not replaced the simpler

part II *General Engineering Information*

gravity slide method used for most pollen counts. Counting techniques vary, but daily pollen counts reported in local newspapers during hay fever season usually represent the number of grains found on 180mm² of a 24h gravity slide.

Hay fever sufferers may experience the first symptoms when the pollen count is 10 to 25; in some localities, maximum figures for the seasonal peak may approach 1000 or more for a 24h period, depending on the sampling and reporting methods used by the laboratory. Translation of gravity counts by special formulas to a volumetric basis (i.e., number of grains per unit volume of air) is unreliable, due to the complexity of the modifying factors. When such information is important, it should be obtained directly by a volumetric instrument. The number of pollen grains per cubic yard of air varies from 2 to 20 times the number found on 1cm² of a 24 h gravity slide, depending on grain diameter, shape, specific gravity, wind velocity, humidity, and physical placement of the collecting plate.

Pollen grains can be removed from the air more readily than the dust particles prevalent in outdoor air or those produced by dusty processes, since a larger fraction of them are in the size range easily removed by building HVAC filters.

Whole-grain pollens are easily removed from the outside air entering a ventilation system with medium-efficiency [35 to 45% (MERV 9 to 11)] filters selected to remove 99% of particles 10μm and greater. Once they have entered a building, whole-grain pollens settle rapidly, reducing concentrations without the need for air cleaning. On resuspension from occupant activities, the wholegrain pollens may disintegrate into fragments, which may possibly be controlled effectively with a high-efficiency [70 to 95% (MERV 12 to 14)] filter capable of removing a high percentage of particles in the 0.3 to 3.0μm range.

Sampling for Bioaerosols

Sampling for biological agents such as fungi and bacteria can include visual observation of colonies, collection of bulk or surface samples, or air sampling. The principles of sampling and analysis for microorganisms are reviewed by Chatigny (1983). ACGIH (1989) and AIHA (1996) have developed assessment guidelines for the collection of microbiological particulates. An introduction is given below.

Preassessment. Sampling for microorganisms should be undertaken when medical evidence indicates the occurrence of diseases such as humidifier fever, hypersensitivity pneumonitis, allergic asthma, and allergic rhinitis. A walk-through examination of the indoor environment for visual detection of possible microbial reservoirs and amplification sites should be performed before sampling. Note that a visual examination will miss reservoirs that are behind walls. If a reservoir or amplifier is visually identified, it is useful to obtain bulk or source samples from it. Also, removal of clearly identified reservoirs and amplifiers is preferable to complicated and costly air-sampling procedures.

Air Sampling. The same principles that affect the collection of an inert particulate aerosol also govern air sampling for microorganisms. Air sampling is not likely to yield useful data and information unless the sample collected is representative of exposure. The most representative samples are those that are collected in breathing zones over the range of aerosol concentrations. Presently, no personal sampling method has been proposed that is sensitive enough for any bioaerosol (Burge 1995). Thus, ambient sampling designs that obtain reasonable estimates of exposures of given populations over representative periods are necessary. Because *Legionella* requires special nutrients for growth and does not produce resistant spores, this bacterium is difficult to recover from air.

Concentrations of microorganisms in the atmosphere vary from a few to several hundred per cubic metre, depending on many factors. The sampling method for microorganisms has an effect on the measured count. Collection on **dry filter paper** can cause count degradation because of the dehydration loss of some organisms. **Glass impingers** may give high counts because agitation can cause clusters to break up into smaller individual organisms. **Slit samples** may give a more accurate colony count.

The viability and/or antigenicity of the microbial particulates must be protected during sampling. In general, **culture plate impactors**, including multiple-and single-stage devices as well as **slit-to-agar samplers**, are most useful in office environments where low concentrations of bacteria and fungi are expected. Because not all microorganisms will grow on the same media, liquid impingement subculturing may be more suitable. **Filter cassette samplers** are useful for microorganisms or components of microorganisms (i.e., endotoxins), although binding to glass and plastic has been reported (Milton et al. 1990). Filter cassettes can also be used for spore counts. Area sampling is often used. Some investigators attempt to replicate exposure conditions through disturbance of the environment (semiaggressive sampling) such as occurs through walking on carpets, slamming doors, and opening books or file cabinets.

Viruses, many bacteria, algae, and protozoa are more difficult to culture than fungi, and air-sampling methodology for these agents is less well known and defined (ASTM 1990).

Data Interpretation. Rank order assessment is a method used to interpret air-sampling data for microorganisms (ACGIH 1989). Individual organisms are listed in descending order of abundance for a complainant indoor site and for one or more control locations. The predominance of one or more microbes in the complainant site, but not in the control sites or outdoors, suggests the presence of an amplifier for that organism. In the example in Table 3, *Tritirachium* and *Aspergillus* were the predominant fungi represented in complainant locations in an office building, where *Cladosporium* and *Fusarium* dominated outdoor collections. In this case, *Tritirachium* and *Aspergillus* were being amplified in the building. In addition to comparing individual organisms, indoor-outdoor ratios of overall quantities of culturable microorganisms are useful.

part Ⅱ *General Engineering Information*

Example Case of Airborne Fungi in Building and in Outdoor Air   Table 3

Location	cfu/m³	Rank Order Taxa
Outdoors	210	*Cladosporium* > *Fusarium* > *Epicoccum* > *Aspergillus*
Complainant Office #1	2500	*Tritirachium* > *Aspergillus* > *Cladosporium*
Complainant Office #2	3000	*Tritirachium* > *Aspergillus* > *Cladosporium*

Notes:

cfu/m³ = Colony-forming units per cubic metre of air.

Culture media was malt extract agar (ACGIH 1989).

Control of Bioaerosols

When maximum removal of airborne microorganisms is either necessary or desirable, high-efficiency particulate air (HEPA) or ultralow penetration air (ULPA) filters are used. These filters create essentially sterile atmospheres and are more frequently employed than chemical scrubbers and ultraviolet radiation for control of airborne microorganisms. They have been used to prevent cross-infection in hospitals, to protect clean rooms from contamination, and to assemble and launch space probes under sterile conditions.

In many situations, total control of airborne microorganisms is not required. For these applications, there are various other types of high-efficiency dry media extended-surface filters that will provide the necessary efficiency. These filters have lower pressure differentials than HEPA filters operating at the same face velocity and, when properly selected, will remove the contaminants of concern.

WORDS AND EXPRESSIONS

activated carbon	活性碳
aerosol	*n.* 气溶胶
airborne	*adj.* 空气中的，空气传播的，气载的
algae	*n.* 藻类，海藻
alkaline	*adj.* 碱的，碱性的
allergic	*adj.* 过敏的，患过敏症的
ammonium	*n.* 铵
amoebae	*n.* 阿米巴虫，变形虫
anthropogenic	*adj.* 人为的，人类造成的
antibiotic	*adj.* 抗生的，抗菌的
antigenicity	*n.* 抗原性
aquatic	*adj.* 水生的，水栖的
aspergillus	曲霉菌
asphyxiant	*n.* 窒息剂

association	群丛，在一特定地理区域里由一种或两种主要种类构成的的大量有机体
asthma	n. 哮喘
attractor plate	集尘板
backdrafting	逆气流的
benzene	n. 苯
bimodal distribution	双峰分布
bioaerosol	n. 生物气溶胶
bioeffluent	生物污染物
blind end	盲端
botanical	adj. 植物学的
bracket fungi	檐状菌
bromine	n. 溴
Brownian motion	布朗运动
butane	n. 丁烷
carbonaceous	adj. 碳的，碳质的，含碳的
cascade impactor	冲击采样器
cataract	n. 白内障
chlorine	n. 氯
cinder	n. 煤渣，灰烬
colony	n. （生物）群体
colony-forming unit	菌落数
colorimetric	adj. 比色的
condensation nucleus counter	凝结核计数器（CNC）
cross-infection	交叉感染
crustal material	地壳物质
cryogenic	adj. 低温的
culture media	培养基
culture mix	混合培养
culture plate impactor	培养皿撞击采样器
dander	n. 鳞屑
data logger	数据记录仪
degreaser	去（油）污剂，脱脂剂
dehydration	n. 脱水
desorption	n. 解吸附作用
discoloration	n. 变色，退色
electrostatic precipitator	静电收尘器
embrittlement	n. 脆化，变脆
endotoxin	n. 内毒素（由某些特定细菌产生并在该细菌细胞裂变

	时释放出来的一种毒素)
expander	n. 发泡剂
exposure dose	暴露剂量
filter cassette sampler	过滤盒采样器
fluorescence	n. 荧光
fluorine	n. 氟
formaldehyde	n. 甲醛
fume	n. 凝结固体烟雾
fungal spore	真菌孢子
habitat	n. (动植物的)生活环境、栖息地
hay fever	n. 花粉热,干草热
haze	n. 薄雾,阴霾
hexane	n. 乙烷
host	n. 宿主,寄主,有另一种生物寄生其上的动植物
hydrochloric acid	盐酸
hydrogen sulfide	氢化硫
hygroscopic	adj. 吸湿的
hypersensitivity	n. 超敏性
impinger	撞击采样器
inhalable	adj. 可吸入的
ionization	n. 离子化,电离
irritating	adj. 刺激的
legionella pneumophila	军团菌
light field microscope	明视野显微镜
light-scattering	光散射
macroscopic	adj. 肉眼可见的
malignant melanoma	n. 恶性脓包,皮肤炭疽
malt extract agar	麦芽膏琼脂
microscopic	adj. 用显微镜可见的
mildew	n. 霉,霉菌
minimum efficiency reporting value (MERV)	最小效率报告值
mist	n. 霭
mite	n. 螨虫
mold	n. 霉菌
nasal	adj. 鼻的
nitrate	n. 硝酸盐,硝酸钾
nominal sieve opening	名义筛网孔眼
nutrient	n. 营养物

oil immersion objective		浸油物镜
overabundance	n.	过多
pesticide	n.	杀虫剂
photosynthetic	adj.	促进光合作用的
plasticizer	n.	可塑剂
pneumonitis	n.	肺炎
propellant	n.	挥发剂，发射剂
protozoa	n.	原生动物，原形动物
puffball	n.	马勃（菌）
radon	n.	氡
reagent	n.	试剂
replicate	v.	复制
reproduce	v.	繁殖，再生
respirable	adj.	可呼吸的
rhinitis	n.	鼻炎，鼻黏膜炎
scale	n.	水垢
scrubber	n.	洗涤器
sewer	n.	下水道
slit sample		缝隙采样法
slit-to-agar sampler		缝隙式琼脂培养基采样器
soot	n.	煤烟，烟灰
sorbent	n.	吸附剂

NOTATIONS

(1) The major classes of air contaminants are particulate and gaseous. The **particulate** class covers a vast range of particle sizes from dust large enough to be visible to the eye to submicroscopic particles that elude most filters. Particulates may be solid or liquid.

空气污染物的主要分类是粒状污染物和气体状污染物。粒状污染物包括的粒径范围很广，从大到可见的粉尘到大部分过滤器都无法过滤的亚显微粒子。粒状污染物可以是固态或液态。

(2) Smog is often associated with temperature inversions in the atmosphere that prevent normal dispersion of contaminants.

烟雾通常与气温逆增有关，妨碍正常的污染物扩散。

(3) Most particles are irregular in shape, and it is useful to characterize their size in terms of some standard particle. For example, the **aerodynamic (equivalent) diameter** of a particle is defined as the diameter of a unit-density sphere having the same gravitational settling velocity as the particle in question.

大多数微粒的形状都是不规则的，以某种标准颗粒来描述它们的大小是有用的。例如，某一微粒的容气动力(当量)直径则定义为具有与之相同重力沉降速度的单位密度球的直径作为该微粒的直径。

(4) The particle size distribution in any sample can be expressed as the percentage of the number of particles smaller than a specified size, area, or mass.

在任一采样中，粒径分布可以表示为小于某一规定粒径、面积或质量粒数的百分数。

(5) Thus, for particles $10\mu m$ and larger, 4% by mass, a 90% arrestance filter essentially removes 90% of 4% for an efficiency by mass of 3.6%.

这样，大于等于$10\mu m$的微粒按质量占4%，一台90%捕集率的过滤器实质上除去4%的90%，即3.6%的质量效率。

(6) This standard defines a laboratory method for assessing the performance of media filters using an optical particle counter to measure particle counts upstream and downstream of the filter in 12 size ranges between 0.3 and $10\mu m$.

该标准规定了一种评价中效过滤器性能的实验方法，即用一台光学粒子计数器在0.3到$10\mu m$之间以12个粒径间隔测量过滤器上、下游的粒子计数。

(7) Another indirect method measures the **optical density** of the collected dust based on the projected area of the particles. Dust particles can be sized with graduated scales or optical comparisons using a standard **microscope**.

另一种间接的方法是根据微粒的投影面积测定收集粉尘的光学密度。粉尘微粒可以用分度尺或用标准显微镜的光学比较来测定大小。

(8) The lower limit for sizing with the light field microscope is approximately $0.9\mu m$, depending on the vision of the observer, the dust color, and the contrast available. This size can be reduced to about $0.4\mu m$ by using oil immersion objective techniques. Dark field microscopic techniques reveal particles smaller than these, to a limit of approximately $0.1\mu m$. Smaller submicroscopic dusts can be sized and compared with the aid of an electron microscope.

用视野显微镜来定微粒尺寸的下限约为$0.9\mu m$，取决于观察者的视力、粉尘颜色和可达到的对比度。通过采用浸油物镜技术，该尺寸可以减小到约$0.4\mu m$。暗视野显微技术能显示比这更小的微粒，到约$0.1\mu m$。更小的亚显微粉尘可用电子显微镜来定大小和比较。

(9) Translation of gravity counts by special formulas to a volumetric basis (i.e., number of grains per unit volume of air) is unreliable, due to the complexity of the modifying factors.

由于修正系数的复杂性，用专门的公式将重力计数转换为体积基数(即每单位体积空气所含颗粒个数)是不可靠的。

(10) Rank order assessment is a method used to interpret air-sampling data for microorganisms. Individual organisms are listed in descending order of abundance for a complainant indoor site and for one or more control locations.

排序评估法是一种用来诠释微生物的空气采样数据的方法。各个生物体按抱怨者的室内位置和一个或多个控制点的数量降序列出。

(11) 注意下列句子中下划线短语或词语的用法：

1) For example, particles 0.1μm or less in diameter (but still above electron microscope minimum detection size of about 0.005 μm) make up(组成、构成)80% of the number of particles in the atmosphere but contribute only 1% of the mass.

2) Suspended particles in urban air are predominantly smaller than 1μm (aerodynamic diameter) and have a distribution that is approximately(近似地,大约)log-normal.

3) Suitable methods for determining the quantity of particulate matter in the air vary depending on(依赖于)the amount present and on the size of particles involved.

4) This method is widely used in industrial workplaces, where there may be significant numbers of large particles, but is not sensitive enough(不足于) for evaluating office environments.

5) Other sizing techniques may take into account(考虑)velocity of samplings in calibrated devices and actual settlement measurements in laboratory equipment.

6) For example, a microscopic technique may measure longest dimension, while impactor results are based on(基于)aerodynamic behavior.

7) If smaller particles detectable by other means, such as an electron microscope or condensation nucleus counter, are also included, the total particle concentration would be greater than the above concentrations by a factor of 10 to 100(比…大10倍到100倍).

8) Bioaerosols are airborne microbiological particulate matter derived from(由…衍生而来) fungi, bacteria, viruses, protozoa, algae, pollen, mites, and their cellular or cell mass components.

9) Problems of concern(关心的,关注的) to engineers occur when microorganisms grow and reproduce indoors.

10) Microorganisms break down(分解)complex molecules found in dead organic materials to simple substances such as carbon dioxide, water, and nitrates.

11) Public interest has focused on(集中于)airborne microorganisms responsible for(是造成…的(主要)原因,对…负责)diseases and infections, primarily bacteria and viruses.

12) A variety of airborne microorganisms are of economic significance(重要的) and can cause product contamination or loss.

13) Thus, they cannot increase in number unless(不…如果不) supplied with a suitable food source such as small quantities of dust, paper, or wood.

14) Other airborne bacteria, especially within closed occupied spaces, may originate from(发源于) droplet nuclei caused by actions such as sneezing or be carried on human or animal skin scales.

15) Pollen grains discharged by weeds, grasses, and trees and capable of causing hay fever have properties of special interest(有兴趣的) to air-cleaning equipment designers.

16) However, despite(不管,尽管)their greater accuracy, these methods have not replaced the simpler gravity slide method used for most pollen counts.

part II *General Engineering Information*

EXERCISES

(1) 解释下列句子中下划线词语的含义，并翻译整个句子。

1) At atmospheric pressure, oxygen concentrations less than 12% or carbon dioxide concentrations greater than 5% are dangerous, even for short periods. Lesser deviations from normal composition can be hazardous under prolonged <u>exposures</u>.

2) Fine particles are usually more chemically complex, anthropogenic, and <u>secondary</u> in origin, while the coarse particles are predominantly <u>primary</u>, natural, and chemically inert.

3) Although most of the mass is contributed by intermediate and larger particles, over 80% of the area (staining) contamination is supplied by particles less than 1μm in diameter, which is in the center of the <u>respirable particle size range</u> and is the size most likely to remain in the lungs.

4) Growth of microbial <u>populations</u> to excessive concentrations is generally associated with inadequate preventive maintenance of these systems, at least in cooling towers.

5) The <u>survival</u> of *Legionella* is enhanced by a variety of parameters including but <u>not limited to</u> warm temperatures, particular algal and protozoan associations, and symbiotic relationships with certain aquatic plants.

6) On <u>resuspension</u> from occupant activities, the wholegrain pollens may <u>disintegrate</u> into fragments, which may possibly be controlled effectively with a high-efficiency [70 to 95% (MERV 12 to 14)] filter capable of removing a high percentage of particles in the 0.3 to 3.0μm range.

7) The most representative samples are those that are collected in <u>breathing zones</u> over the range of aerosol concentrations.

8) Because *Legionella* requires special nutrients for growth and does not produce resistant spores, this bacterium is difficult to <u>recover</u> from air.

(2) 请用英语表达。

可呼吸尘，可吸入尘，悬浮颗粒物

烟，雾，烟雾，霭，气溶胶

光学粒子计数器，凝结核计数器，冲击采样器，静电收尘器

(有害物)容许最高浓度

Part III

HVAC systems and equipment

LESSON 8

CENTRAL COOLING AND HEATING

THE energy used to heat and cool many buildings often comes from a central location in the facility. The energy input may be any combination of electricity, oil, gas, coal, solar, geothermal, etc. This energy is typically converted into hot or chilled water or steam that is distributed throughout the facility for heating and air conditioning (cooling). Centralizing this function keeps the conversion equipment in one location and distributes the heating and cooling in a more readily usable form. Also, a central cooling and heating plant provides higher diversity and generally operates more efficiently with lower maintenance and labor costs than a decentralized plant. However it does require space at a central location and a potentially large distribution system.

This lesson addresses the design alternatives that should be considered when centralizing the cooling and heating sources in a facility.

The economics of these systems require extensive analysis. Boilers; gas, steam, and electric turbine-driven centrifugal refrigeration units; and absorption chillers may be installed in combination in one plant. In large buildings with core areas that require cooling and perimeter areas require heating, one of several heat reclaim systems may heat the perimeter to save energy.

The choice of equipment for a central heating and cooling plant depends on the following:
- Required capacity and type of use
- Cost and types of energy available
- Location and space for the equipment room
- A stack for any exhaust gases and its environmental impact
- Safety requirements for the mechanical room and surrounding areas
- Type of air distribution system
- Owning and operating costs

Many electric utilities impose very high charges for peak power use or, alternatively, offer incentives for off-peak use. This policy has renewed interest in both water and ice thermal storage. The storage capacity installed for summer load leveling may also be available for use in the winter, making heat reclaim a more viable option. With ice storage, the low-temperature ice water can provide colder air than that available from a

---

Extracted from Chapter 4 of the ASHRAE *Handbook-HVAC Systems and Equipment*.

conventional system that supplies air at 10 to 13℃. The use of a higher water temperature rise and a lower supply air temperature, lowers the required pumping and fan energy and, in some instances, offsets the energy penalty due to the lower temperature required to make ice.

REFRIGERATION EQUIPMENT

The major types of refrigeration equipment in large systems use reciprocating compressors, helical rotary compressors, centrifugal compressors, and absorption chillers.

These large compressors are driven by electric motors, gas and diesel engines, and gas and steam turbines. The compressors may be purchased as part of a refrigeration chiller that includes the compressor, drive, chiller, condenser, and necessary safety and operating controls. Reciprocating and helical rotary compressor units are frequently field assembled, and include air-cooled or evaporative condensers arranged for remote installation. Centrifugal compressors are usually included in packaged chillers.

Absorption chillers are water-cooled. They use a lithium bromide/water or water/ammonia cycle and are generally available in the following four configurations: (1) direct fired, (2) indirect generated by low-pressure steam or hot water, (3) indirect generated by high-pressure steam or hot water, and (4) indirect generated by hot exhaust gas. Small direct-fired chillers are single-effect machines with capacities of 12 to 90kW. Larger, direct-fired, double-effect chillers in the 350 to 7000kW range are also available.

Low-pressure steam at 100kPa or low temperature hot water heats the generator of single-effect absorption chillers with capacities from 180 to 5600kW. Double-effect machines use high-pressure steam up to 1000kPa or high-temperature hot water at an equivalent temperature. Absorption chillers of this type are available from 1200 to 7000kW.

In large installations the absorption chiller is sometimes combined with steam turbine-driven centrifugal compressors. Steam from the noncondensing turbine is piped to the generator of the absorption machine. When a centrifugal unit is driven by a gas turbine or an engine, an absorption machine generator may be fed with steam or hot water from the engine jacket. A heat exchanger that transfers the heat of the exhaust gases to a fluid medium may increase the cycle efficiency.

Cooling Towers

Water from condensers is usually cooled by the atmosphere. Either natural draft or mechanical draft cooling towers or spray ponds are used to reject the heat to the atmosphere. Of these, the mechanical draft tower, which may be of the forced draft, induced draft, or ejector type, can be designed for most conditions because it does not depend on the wind. Air-conditioning systems use towers ranging from small package units of 20 to 1800kW to field-erected towers with multiple cells in unlimited sizes.

If the cooling tower is on the ground, it should be at least 30m away from the building for two reasons: (1) to reduce tower noise in the building; and (2) to keep discharge air from fogging the building's windows. Towers should be kept the same distance from parking lots to avoid staining car finishes with water treatment chemicals. When the tower is on the roof, its vibration and noise must be isolated from the building. Some towers are less noisy than others, and some have attenuation housings to reduce noise levels. These options should be explored before selecting a tower. Adequate room for airflow inside screens should be provided to prevent recirculation.

The bottom of many towers, especially larger ones, must be set on a steel frame 1 to 1.5m above the roof to allow room for piping and proper tower and roof maintenance. Pumps below the tower should be designed for an adequate net positive suction pressure, but they must be installed to prevent draining the piping on shutdown.

The tower must be winterized if required for cooling at outdoor temperatures below 2℃. Winterizing includes the ability to bypass water directly into the tower basin or return line (either automatically or manually, depending on the installation) and to heat the tower pan water to a temperature above freezing. Heat may be added by steam or hot water coils or by electric resistance heaters in the tower pan. Also, an electric heating cable on the condenser water and makeup water pipes and insulation on these heat-traced sections is needed to keep the pipes from freezing. Special controls are also required when a cooling tower is operated near freezing conditions. Where the cooling tower will not operate in freezing weather, provisions for draining the tower and piping must be made. Draining is the most effective way to prevent the tower and piping from freezing.

Careful attention must also be given to water treatment to minimize the maintenance required by the cooling tower and refrigeration machine absorbers and/or condenser.

Cooling towers may also cool the building in off-seasons by filtering and directly circulating the condenser water through the chilled water circuit, by cooling the chilled water in a separate heat exchanger, or by using the heat exchangers in the refrigeration equipment to produce thermal cooling. Towers are usually selected in multiples so that they may be run at reduced capacity and shut down for maintenance in cool weather. Chapter 36 includes further design and application details.

Air-Cooled Condensers

Air-cooled condensers pass outdoor air over a dry coil to condense the refrigerant. This process results in a higher condensing temperature and, thus, a larger power input at peak condition; however, over 24 hours this peak time may be relatively short. The air-cooled condenser is popular in small reciprocating systems because of its low maintenance requirements.

Evaporative Condensers

Evaporative condensers pass outdoor air over coils sprayed with water, thus taking advantage of adiabatic saturation to lower the condensing temperature. As with the cooling tower, freeze prevention and close control of water treatment are required for successful operation. The lower power consumption of the refrigeration system and the much smaller footprint of the evaporative versus the air-cooled condenser are gained at the expense of the cost of the water used and increased maintenance cost.

HEATING EQUIPMENT

A boiler is the most common device used to add heat to the working medium, which is then distributed throughout the facility. Although steam is an acceptable medium for transferring heat between buildings or within a building, low-temperature hot water provides the most common and more uniform means of perimeter and general space heating. The working medium may be either water or steam, which can further be classified by its temperature and pressure range. The term hot-water boiler applies to fuel-fired units that heat water for heating systems. Water heaters differ in that they usually do not have enough space in the top section for use as a steam boiler, but in many respects a water heating boiler is the same as a steam heating boiler of the same type of construction. Many steam heating boilers may serve as water heaters if properly arranged, fitted, and installed. Steam and hot water boilers use gas, oil, coal, electricity, and sometimes, waste material for fuel.

Fuel Section

Functionally a boiler has two major sections. For all fuel burning boilers, that fuel must first be burned in the combustion section of the boiler. Control of both the fuel and air greatly effect the efficiency of the combustion process. This is also the primary point at which the heating capacity of the boiler is regulated. On smaller boilers that only turn on or off, the air/fuel ratio is fixed and is not actively controlled. A next step of regulation may have one or two intermediate steps of fuel input, such a low fire-high fire, with the air-fuel ratio fixed at each level of fuel input. A larger or proportionally regulated boiler may have its fuel flow regulated to any level. Then a complimentary air regulation method is also required to ensure sufficient oxygen for efficient fuel combustion without either heating excess air that is wasted up the exhaust stack, or creating hazardous or polluting compounds.

Heat Transfer Section

The second major section of the boiler is the heat transfer section where heat from the combustion process is transferred to the working medium that will distribute the heat. The effectiveness of these heat transfer surfaces depends on the temperature difference between the fluid, water, or steam on one side and the gases on the other side, and on the circulation rate of both the water and the hot gases. Heating surfaces are classified as direct and indirect. Direct surfaces are those upon which the light of the fire is present; they are very effective in transferring heat to the boiler water because the high furnace temperatures promote heat transfer both by radiation and convection. Indirect heating surfaces are those in contact with flue gases only and are progressively less effective from the standpoint of heat transfer as the flue gases become cooled. Boiler manufactures generally apply a combination of these heat transfer surfaces for boilers. In **fire tube boilers** the products of combustion flow inside tubes and the transfer medium (water) flows outside. In **water tube boilers** the products of combustion flow outside tubes and the transfer medium flows inside the tubes. **Cast iron sectional boilers** transfer the heat from the combustion gases to the water through cast iron sections.

Working Pressure

Pressures in the boiler also affect its classification. The combustion section can be regulated at a slightly negative pressure relative to the atmosphere to ensure that the combustion gases do not flow anywhere except up the stack. In a **natural draft boiler**, the chimney effect of the stack draws the combustion products up the stack. Alternately, the boiler may use a fan to force air through the burner and boiler in a **forced draft** arrangement. In this case the combustion chamber and boiler itself are at a positive pressure relative to the surrounding atmosphere.

The pressure of the working medium (steam or water) can be low, medium, or high. For hot water boilers more so then steam boilers the location in the facility as well as its elevation relative to the rest of the facility impacts the working pressure. Pumping in or out of a water boiler can effect its operating/dynamic pressure; and the boiler's location, either in a basement or on a roof, can effect its standby/static pressure.

DISTRIBUTION

The steam or hot water is the working medium of the heating system that transfers the heat produced by the boiler to the areas where it will be used. Steam typically is distributed by its inherent pressure; but once it is condensed, it must rely on gravity or pumps to return to the boiler. The issue of condensate traps and the return is one of the important design and operational concerns for steam. If water is the working medium, pumps distribute it from the boiler.

Pumps

HVAC pumps are usually centrifugal pumps. Pumps for large, heavy-duty systems have a horizontal split case with a double-suction impeller for easier maintenance and higher efficiency. End suction pumps, either close coupled or flexible connected, may be used for smaller tasks. Pumps can be installed in-line or be base mounted.

Pumps are used to distribute or circulate the following fluids:

Part III　HVAC systems and equipment

- Primary and secondary chilled water
- Primary and secondary hot water
- Condenser water
- Condensate
- Boiler feed
- Fuel oil

When pumps handle hot liquids or have a high inlet pressure drop, the required net positive suction pressure must not exceed the net positive suction pressure available at the pump. It is common practice to provide spare pumps to maintain continuity in case of a pump failure. The chilled water and condenser water system characteristics may permit using one spare pump for both systems, if they are properly valved and connected to both manifolds.

Piping

HVAC piping systems can be divided into two parts—the piping in the main equipment room and the piping required to deliver chilled or hot water/steam to the air-handling equipment throughout the building.

The major piping in the equipment room includes fuel lines, refrigerant piping, and steam and water connections.

Chilled water systems may be designed for constant flow to cooling coils with three-way valves or to coils with balancing valves. Because flow is constant the load on the system is proportional to the temperature difference across the main supply and return lines.

Multiple chillers on constant flow systems are generally piped in series. Piping the chillers in parallel is not recommended because under light load conditions it is difficult to maintain the design supply water temperature when return water through the inactive chiller(s) is mixing with cold water from the active chiller(s).

Very few constant flow systems are designed today because of the increased cost to continuously circulate the design flow and the difficulty in controlling the supply water temperature.

Variable flow chilled water systems, serve cooling coils with two-way modulating control valves so that the flow is proportional to the coil load. In order to maintain a constant flow through the online chiller(s), a primary-secondary pumping arrangement is generally used.

The thermal load is proportional to the flow times the temperature difference across the secondary supply and return mains. For this reason both flow and temperature must be measured accurately to determine the load on the system. An accurate measurement of the load is essential in order to operate the correct number of chillers and/or boilers to optimize the system for peak efficiency. The temperature sensors should be matched to 0.1℃ and the flow sensor should be accurate within 1% of the range.

The crossover line on primary-secondary systems may be sized for the flow through the largest chiller rather than the total flow of all chillers.

INSTRUMENTATION

All equipment must have adequate pressure gages, thermometers, flowmeters, balancing devices, and dampers for effective performance, monitoring, and commissioning. In addition, capped thermometer wells, gage cocks, capped duct openings, and volume dampers should be installed at strategic points for system balancing.

Microprocessors and the related software to establish control sequences are replacing pneumatic and electric relay logic. Output signals are converted to pneumatic or electric commands to actuate HVAC hardware.

A central control console to monitor the many system points and overall system performance should be considered for any large, complex air-conditioning system. A control panel permits a single operator to monitor and perform functions at any point in the building to increase occupant comfort and to free maintenance staff for other duties.

The coefficient of performance for the entire chilled water plant can now be monitored and allows the chilled water plant operator to determine the overall operating efficiency of a plant.

All instrument operations where heat or cooling output are measured should have instrumentation whose calibration is traceable to NIST, the National Institute of Standards and Technology.

SPACE REQUIREMENTS

In the initial phases of building design, the engineer seldom has sufficient information to render the design. Therefore, most experienced engineers have developed rules of thumb to estimate the building space needed. The air-conditioning system selected, the building configuration, and other variables govern the space required for the mechanical system. The final design is usually a compromise between what the engineer recommends and what the architect can accommodate. The design engineer should keep the owner and facility engineer informed, whenever possible, about the HVAC analysis and system selection. Space criteria should satisfy both the architect and the owner or the owner's representative. Where designers cannot negotiate a space large enough to suit the installation, the judgement of the owners should be called upon.

Although few buildings are identical in design and concept, some basic criteria apply to most building and help allocate space that approximates the final requirements. These space requirements are often expressed as a percentage of the total building floor area. For example, the total mechanical and electrical space requirements range from 4 to 9% of the gross building area, with most buildings falling within the 6 to 9% range.

The arrangement and strategic location of the mechanical spaces during planning impact the percentage of space required. The relationship of outdoor air intakes with respect to loading docks, exhaust, and other contaminating sources should be considered during architectural planning.

The main electrical transformer and switchgear rooms should be located as close to the incoming electrical service as practical. If there is an emergency generator, it should be located considering (1) proximity to emergency electrical loads, (2) sources of combustion and cooling air, (3) fuel sources, (4) ease of properly venting exhaust gases to the outdoors, and (5) provisions for noise control.

The main plumbing equipment usually contains gas and domestic water meters, the domestic hot water system, the fire protection system, and such elements as compressed air, special gases, and vacuum, ejector, and sump pumps. Some water and gas utilities require a remote outdoor meter location.

The heating and air-conditioning equipment room houses (1) the boiler, or pressure-reducing station, or both; (2) the refrigeration machines, including the chilled water and condensing water pumps; (3) converters for furnishing hot or cold water for air conditioning; control air compressors; (4) vacuum and condensate pumps; and (5) miscellaneous equipment. For both chillers and boilers, especially in a centralized application, full access is needed on all sides for extended operation, maintenance, annual inspections of tubes, and tube replacement and repair. Local codes and ASHRAE *Standard* 15 should be consulted for special equipment room requirements. Many local jurisdictions require separation of refrigeration and fuel fired equipment.

It is often economical to locate the refrigeration plant at the top of the building, on the roof, or on intermediate floors. These locations closer to the load allow the equipment to operate at a lower pressure. The electrical service and structural costs will be greater, but these may be offset because the energy consumption and the condenser and chilled water piping cost may be less. The boiler plant may also be placed on the roof, which eliminates the need for a chimney through the building.

CENTRAL PLANT LOADS

The design cooling and heating loads are determined by considering the entire portion or block of the building served by the air-and-water system at the same time. Because the load on the secondary water system depends on the simultaneous demand of all spaces, the sum of the individual room or zone peaks is not considered.

The peak cooling load time is influenced by the outdoor wet-bulb temperature, the period of building occupancy, and the relative amounts of the east, south, and west exposures. Where the magnitude of the solar load is about equal for each of the above exposures, the building peak usually occurs on a midsummer afternoon when the west solar load and outdoor wet-bulb temperature are at or near concurrent maximums.

If the side exposed to the sun or interior zone loads require chilled water in cold weather, the use of condenser water with a water-to-water heat exchanger should be considered. Varying refrigeration loads require the water chiller to operate satisfactorily under all conditions.

If water supply temperature or quantities are to be reset at times other than at peak load, the adjusted settings must be adequate for the most heavily loaded space in the building. An analysis of individual room load variations is required.

The peak heating load may occur when the facility must be warmed to occupancy conditions after an unoccupied weekend setback. The peak demand may also occur during unoccupied conditions when the ambient environment is at its harshest and there is little internal heat gain to assist the heating system. Another possible maximum may occur during occupied times if significant outdoor air must be preconditioned or some other process (maybe non-HVAC) requires significant heat. The designer must analyze how the facility will be used.

WORDS AND EXPRESSIONS

absorber	n. 吸收器
absorption chiller	吸收式冷水机
attenuation housing	减噪房
boiler feed	锅炉补水
calibration	n. 校正
cast iron sectional boiler	铸铁段锅炉
centrifugal refrigeration units	离心式制冷机组
circulation rate	循环流量
commissioning	调试
complimentary	另人满意的
condensate	凝结水
console	n. [计] 控制台
crossover line	交叉管
decentralized plant	分散式设备
direct fired	直燃式
diversity	n. 多样性
double-suction impeller	双吸式叶轮
ejector	n. 喷射器，喷射泵
emergency generator	应急发电机
end suction pump	单吸水泵
engine jacket	发动机夹套
evaporative condenser	蒸发式冷凝器
exhaust gases	废气

facility	*n.*	设施
forced draft		强制通风
geothermal	*adj.*	地热的
heavy-duty		大负荷的
helical rotary compressor		螺杆式压缩机
induced draft		诱导通风
intermediate	*adj.*	中间的，过渡的
jurisdiction	*n.*	权限，管辖区域
labor costs		劳动力费用
lithium bromide		溴化锂
load leveling		负荷调整（削峰填谷）
maintenance	*n.*	维护
manifolds		多支管
miscellaneous equipment		辅助设备
National Institute of Standards and Technology		美国国家标准和技术学会
off-season		过渡季、淡季（春、秋两季）
packaged chiller		组装式冷水机
perimeter areas		外区
pneumatic	*adj.*	气动的
primary and secondary chilled water		一次冷冻水和二次冷冻水
reciprocating compressor		往复式压缩机
rule of thumb		经验法则
spare pumps		备用水泵
stack		烟囱或烟道
sump pump		污水泵
switchgear	*n.*	开关设备
traceable	*adj.*	可跟踪的
winterize	*vt.*	作过冬准备；使防冻

NOTATIONS

(1) The economics of these systems require extensive analysis. Boilers; gas, steam, and electric turbine-driven centrifugal refrigeration units; and absorption chillers may be installed in combination in one plant. In large buildings with core areas that require cooling and perimeter areas require heating, one of several heat reclaim systems may heat the perimeter to save energy.

这些系统的经济性需要做广泛的分析。锅炉，燃气、蒸汽和电力驱动的离心式制冷机组,以及吸收式冷水机都可以组合安装在一个机房内。在内区要求供冷而周边区需要供热的大型建筑中,几个热回收系统的一个可以加热周边区,以节省能量。

(2) Many electric utilities impose very high charges for peak power use or, alternatively, offer incentives for off-peak use. This policy has renewed interest in both water and ice thermal storage. The storage capacity installed for summer load leveling may also be available for use in the winter, making heat reclaim a more viable option.

很多电力公司都对高峰用电采取高昂的收费,而鼓励低谷用电。这项政策已经引起对水和冰热力蓄存技术的兴趣。为夏季负荷调整配置的蓄存容量也可以在冬季使用,使得热回收成为一个更为可行的选择。

(3) These large compressors are driven by electric motors, gas and diesel engines, and gas and steam turbines.

这些大型的压缩机由电机、燃气和柴油发动机及燃气和蒸汽轮机所驱动。

(4) Water from condensers is usually cooled by the atmosphere. Either natural draft or mechanical draft cooling towers or spray ponds are used to reject the heat to the atmosphere.

从冷凝器出来的水通常被大气冷却。可以采用自然通风或机械通风式冷却塔或用喷水池将热量排到大气中去。

(5) Pumps below the tower should be designed for an adequate net positive suction pressure, but they must be installed to prevent draining the piping on shutdown.

比冷却塔低的水泵应具有足够的净正吸入压力,但是它们必须安装成能防止停机时排空管道。

(6) On smaller boilers that only turn on or off, the air/fuel ratio is fixed and is not actively controlled. A next step of regulation may have one or two intermediate steps of fuel input, such a low fire-high fire, with the air-fuel ratio fixed at each level of fuel input. A larger or proportionally regulated boiler may have its fuel flow regulated to any level.

仅进行开关控制的小型锅炉空气/燃料比是固定的,且不能主动控制。进一步的调节控制可以有一级或两级燃料输入的中间值,这样的低火或高火,在每一燃料输入级空气/燃料比都固定。一种更大型或者比例调节的锅炉可能具有调节到任何水平的燃料流量。

(7) For hot water boilers more so then steam boilers the location in the facility as well as its elevation relative to the rest of the facility impacts the working pressure. Pumping in or out of a water boiler can effect its operating/dynamic pressure; and the boiler's location, either in a basement or on a roof, can effect its standby/static pressure.

锅炉在机房中的位置和它们相对于其他设施的高度会影响工作压力,这种影响热水锅炉比蒸汽锅炉要来的大。热水锅炉的泵进和抽出能够影响其运行/动态压力;而锅炉的位置——位于地下室或屋顶——可影响其待机/静态压力。

(8) The issue of condensate traps and the return is one of the important design and operational concerns for steam.

凝结水分离器(疏水器)的出流和回水是蒸汽系统设计和运行的重要事项之一。

(9) Pumps for large, heavy-duty systems have a horizontal split case with a double-suc-

tion impeller for easier maintenance and higher efficiency. End suction pumps, either close coupled or flexible connected, may be used for smaller tasks. Pumps can be installed in-line or be base mounted.

为便于维护和保持较高效率，大型的负荷大的系统的水泵具有卧式分开的机壳、双吸式叶轮。负荷较小的系统可以用单吸泵，或紧密对接，或柔性连接。水泵可以管道式安装，或者装在基座上。

(10) The temperature sensors should be matched to 0.1℃ and the flow sensor should be accurate within 1% of the range.

温度传感器的精度应达到0.1℃，而流量传感器的精度应在其量程的1%之内。

(11) Where the magnitude of the solar load is about equal for each of the above exposures, the building peak usually occurs on a midsummer afternoon when the west solar load and outdoor wet-bulb temperature are at or near concurrent maximums.

在太阳负荷的量级对以上各朝向(东、南、西)差不多的场合，建筑的峰值负荷往往发生在某一个仲夏的下午——西面的太阳和室外湿球温度都同时处于或接近最大值时。

(12) The peak heating load may occur when the facility must be warmed to occupancy conditions after an unoccupied weekend setback.

峰值供热负荷可能发生在某一个无人的周末停机之后，整个设施要被加热到有人的状态。

EXERCISES

(1) 请翻译下列句子。

1) The use of a higher water temperature rise and a lower supply air temperature, lowers the required pumping and fan energy and, in some instances, offsets the energy penalty due to the lower temperature required to make ice.

2) Cooling towers may also cool the building in off-seasons by filtering and directly circulating the condenser water through the chilled water circuit, by cooling the chilled water in a separate heat exchanger, or by using the heat exchangers in the refrigeration equipment to produce thermal cooling.

3) The effectiveness of these heat transfer surfaces depends on the temperature difference between the fluid, water, or steam on one side and the gases on the other side, and on the circulation rate of both the water and the hot gases.

4) Chilled water systems may be designed for constant flow to cooling coils with three-way valves or to coils with balancing valves. Because flow is constant the load on the system is proportional to the temperature difference across the main supply and return lines.

5) In addition, capped thermometer wells, gage cocks, capped duct openings, and volume dampers should be installed at strategic points for system balancing.

(2) 通过阅读文章，请说明 Central Heating and Cooling System 的特点。

LESSON 9

DECENTRALIZED COOLING AND HEATING

PACKAGED unit systems are applied to almost all classes of buildings. They are especially suitable for smaller projects with no central plant where low initial cost and simplified installation are important. These units are installed in office buildings, shopping centers, manufacturing plants, hotels, motels, schools, medical facilities, nursing homes, and other multiple-occupancy dwellings. They are also suited to air conditioning existing buildings with limited life or income potential. Applications also include facilities requiring specialized high performance levels, such as computer rooms and research laboratories.

SYSTEM CHARACTERISTICS

These systems are characterized by several separate air-conditioning units, each with an integral refrigeration cycle. The components are factory designed and assembled into a package that includes fans, filters, heating coil, cooling coil, refrigerant compressor(s), refrigerant-side controls, air-side controls, and condenser. The equipment is manufactured in various configurations to meet a wide range of applications. Examples include window air conditioners, through-the-wall room air conditioners, unitary air conditioners for indoor and outdoor locations, air-source heat pumps, and water-source heat pumps. Specialized packages for computer rooms, hospitals, and classrooms are also available.

Commercial grade unitary equipment packages are available only in pre-established increments of capacity with set performance parameters, such as the sensible heat ratio at a given room condition or the amount of airflow per kilowatt of refrigeration. Components are matched and assembled to achieve specific performance objectives. These limitations make the manufacture of low cost, quality-controlled, factory-tested products practical. For a particular kind and capacity of unit, performance characteristics vary among manufacturers. All characteristics should be carefully assessed to ensure that the equipment performs as needed for the application. Several trade associations have developed standards by which manufacturers may test and rate their equipment.

Extracted from Chapter 5 of the ASHRAE *Handbook-HVAC Systems and Equipment*.

Large commercial/industrial grade equipment can be custom designed by the factory to meet specific design conditions and job requirements. This equipment carries a higher first cost and is not readily available in smaller sizes.

Self-contained units can use multiple compressors to control refrigeration capacity. For variable air volume systems, compressors are turned on or off or unloaded to maintain the discharge air temperature. As airflow is decreased, the temperature of the air leaving the unit can often be reset upward so that a minimum ventilation rate can be maintained. Resetting the discharge air temperature is a way of demand limiting the unit, thus saving energy.

Although the equipment can be applied as a single unit, this chapter covers the application of multiple units to form a complete air-conditioning system for a building. Multiple, packaged-unit systems for perimeter spaces are frequently combined with a central all-air or floor-by-floor system. These combinations can provide better humidity control, air purity, and ventilation than packaged units alone. Air-handling systems may also serve interior building spaces that cannot be conditioned by wall or window-mounted units.

Advantages of Packaged Systems
- Heating and cooling capability can be provided at all times, independent of the mode of operation of other spaces in the building.
- Manufacturer-matched components have certified ratings and performance data.
- Assembly by a manufacturer helps ensure better quality control and reliability.
- Manufacturer instructions and multiple-unit arrangements simplify the installation through repetition of tasks.
- Only one unit conditioner and one zone of temperature control are affected if equipment malfunctions.
- System is readily available.
- One manufacturer is responsible for the final equipment package.
- For improved energy control, equipment serving vacant spaces can be turned off locally or from a central point, without affecting occupied spaces.
- System operation is simple. Trained operators are not required.
- Less mechanical and electrical room space is required than with central systems.
- Initial cost is usually low.
- Equipment can be installed to condition one space at a time as a building is completed, remodeled, or as individual areas are occupied, with favorable initial investment.
- Energy can be metered directly to each tenant.

Disadvantages of Packaged Systems
- Limited performance options may be available because airflow, cooling coil size, and

condenser size are fixed.
- A larger total building installed cooling capacity is usually required because the diversity factors used for moving cooling needs do not apply to dedicated packages.
- Temperature and humidity control may be less stable especially with mechanical cooling at very low loads.
- Standard commercial units are not generally suited for large percentages of outside air or for close humidity control. Custom equipment or special purpose equipment such as packaged units for computer rooms or large custom units may be required.
- Energy use is usually greater than for central systems, if efficiency of the unitary equipment is less than that of the combined central system components.
- Low cost cooling by outdoor air economizers is not always available.
- Air distribution control may be limited.
- Operating sound levels can be high.
- Ventilation capabilities are fixed by equipment design.
- Overall appearance can be unappealing.
- Air filtration options may be limited.
- Discharge temperature varies because control is either on or off or in steps.
- Maintenance may be difficult because of the many pieces of equipment and their location.

ECONOMIZERS

Air-Side Economizers

The air-side economizer takes advantage of cool outdoor air to either assist mechanical cooling or, if the outdoor air is cool enough, provide total cooling. It requires a mixing box designed to allow 100% of the air to be drawn from outside. It can be field-installed accessory that includes an outdoor air damper, relief damper, return air damper, filters, actuator, and linkage. Controls are usually a factory-installed option.

Self-contained units usually do not include return air fans. A variable-volume relief fan must be installed with the air-side economizer. The relief fan is off and discharge dampers are closed when the air-side economizer is inactive.

Advantages of Air-Side Economizers
- Substantially reduces compressor, cooling tower, and condenser water pump energy requirements.
- Has a lower air-side pressure drop than a water-side economizer.
- Reduces tower makeup water and related water treatment.

Disadvantages of Air-Side Economizers
- In systems with larger return air static pressure requirements, return fans or exhaust

fans are needed to properly exhaust building air and intake outside air.
- If the unit's leaving air temperature is also reset up during the airside economizer cycle, humidity control problems may occur and the fan may use more energy.
- Humidification may be required during winter operation.

Water-Side Economizer

The water-side economizer is another option for reducing energy use. ASHRAE *Standard* 90.1 addresses its application. The waterside economizer consists of a water coil located in the self-contained unit upstream of the direct-expansion cooling coil. All economizer control valves, piping between the economizer coil, and the condenser and economizer control wiring can be factory installed (Figure 1).

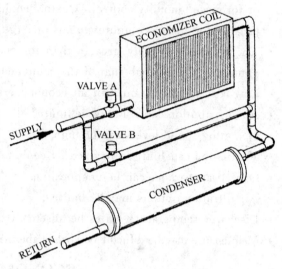

Fig. 1 Water-Side Economizer and Valves

The water-side economizer takes advantage of the low cooling tower or evaporative condenser water temperature to (1) either precool the entering air, (2) assist mechanical cooling, or, (3) if the cooling water is cold enough, provide total system cooling. If the economizer is unable to maintain the supply air set point for variable-air-volume units or zone set point for constant-volume units, factory-mounted controls integrate economizer and compressor operation to meet cooling requirements.

Cooling water flow is controlled by two valves (Figure 1), one at the economizer coil inlet (A) and one in the bypass loop to the condenser (B). Two control methods are common—constant water flow and variable water flow.

Constant water flow control allows constant condenser water flow during unit operation. The two control valves are factory wired for complementary control, where one valve is driven open while the other is driven closed. This keeps water flow through the unit relatively constant.

Variable modulating control allows variable condenser water flow during unit operation. The valve in the bypass loop (B) is an on-off valve and is closed when the economizer is enabled. Water flow through the economizer coil is modulated by valve A, thus allowing variable cooling water flow. As the cooling load increases, valve A opens, increasing water flow through the economizer coil. If the economizer is unable to satisfy the cooling requirements, factory-mounted controls integrate economizer and compressor operation. In this operating mode, valve A is fully open. When the selfcontained unit is not in the cooling mode, both valves are closed. Reducing or eliminating cooling water flow reduces pumping energy.

Advantages of Water-Side Economizers
- Reduces compressor energy by precooling entering air. Often the building load can be completely satisfied with an entering condenser water temperature of less than 13℃. Because the wet-bulb temperature is always less than or equal to the dry-bulb temperature, a lower discharge air temperature is often times available.
- Building humidification may not be required if return air contains sufficient humidity to satisfy the winter requirement.
- No external wall penetration is required for exhaust or outdoor air ducts.
- Mechanical equipment rooms can be centrally located in a building.
- Controls are less complex than for air-side economizers, because they often reside inside the packaged unit.
- Coil can be mechanically cleaned.
- More net usable floor area is available because large fresh air and relief air ducts are unnecessary.

Disadvantages of Water-Side Economizers
- Cooling tower water treatment cost is greater.
- Air-side pressure drop may be increased.
- Condenser water pump may see slightly higher pressure.
- Cooling tower must be designed for winter operation.
- The increased operation(including winter operation)required of the cooling tower may reduce its life.

THROUGH-THE-WALL AND WINDOW-MOUNTED AIR CONDITIONERS AND HEAT PUMPS

A window air conditioner (air-cooled room conditioner) is designed to cool or heat individual room spaces. Window units are used where low initial cost, quick installation, and other operating or performance criteria outweigh the advantages of more sophisticated systems. Room units are also available in through-the-wall sleeve mountings. Sleeve-installed units are popular in low-cost apartments, motels, and homes.

Window units may be used as auxiliaries to a central heating or cooling system or to condition selected spaces when the central system is shut down. In such applications, these units usually serve only part of the spaces conditioned by the basic system. Both the basic system and the window units should be sized to cool the space adequately without the other operating.

A through-the-wall air-cooled room air conditioner is designed to cool or heat individual room spaces. Design and manufacturing parameters vary widely. Specifications range from appliance grade through heavy-duty commercial grade, the latter known as packaged terminal air conditioners (ARI *Standard* 310). With proper maintenance, manufacturers project a life expectancy of 10 to 15 years for these units.

Advantages

- Initial cost is generally less than for a central system adapted to heat or cool each room under the control of the room occupants.
- Because no energy is needed to transfer air or chilled water from mechanical equipment rooms, the energy consumption may be lower than for central systems. However, this advantage may be offset by the lower efficiency of this type of equipment.
- Building space is conserved because ductwork and mechanical rooms are not required.
- Installation is simple. It usually only requires a hole in the wall or displacement of a window to mount the unit and connection to electrical power.
- Generally, the system is well-suited to spaces requiring many zones of individual temperature control.
- Designers can specify electric, hydronic, or steam heat or use an air-to-air heat pump design.
- Service can be quickly restored by replacing a defective chassis with a spare.

Disadvantages

- Equipment life is relatively short, typically 10 years; window units are built to appliance standards, rather than building equipment standards.
- May have relatively high energy use.
- Requires outside air; thus, cannot be used for interior rooms. Packaged terminal air conditioners must be installed on the perimeter of the building.
- The louver and wall box must stop wind-driven rain from collecting in the wall box and leaking into the building. The wall box should drain to the outside.
- Condensate removal can cause dripping on walls, balconies, or sidewalks.
- Temperature control is usually two-position, which causes swings in room temperature.
- Air distribution control is limited.
- Ventilation capabilities are fixed by equipment design.
- Overall appearance can be unappealing.
- Air filtration options are limited.
- Humidification, when required, must be provided by separate equipment.
- Noise levels vary considerably and are not generally suitable for critical applications.
- Routine unit maintenance is required to maintain capacity. Condenser and cooling coils must be cleaned, and filters must be changed regularly.

Design Considerations

Units are usually furnished with individual electric controls. However, when several units are used in a single space, the controls should be interlocked to prevent simultaneous heating and cooling. In commercial applications (e. g. , motels), centrally operated switches can de-energize units in unoccupied rooms.

A through-the-wall or window-mounted air-conditioning unit incorporates a complete air-cooled refrigeration and air-handling system in an individual package. Each room is an individual occupant-controlled zone. Cooled or warmed air is discharged in response to thermostatic control to meet room requirements. The section on Controls describes how controls allow the use of individual room systems during off-schedule hours, but automatically return all systems to normal schedule use.

Each packaged terminal air conditioner has a self-contained, aircooled direct-expansion cooling system; a heating system(electric, hot water, or steam); and controls. Two general configurations are shown—Figure 2 shows a wall box, an outdoor louver, heater section, cooling chassis, and cabinet enclosure; Figure 3 shows a combination wall sleeve cabinet, plus combination heating and cooling chassis with outdoor louver.

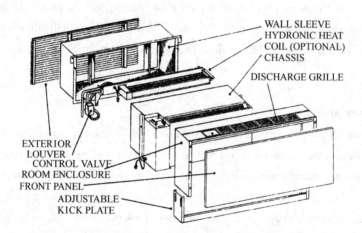

Fig. 2 Packaged Terminal Air Conditioner with Heating
Section Separate from Cooling Chassis

A through-the-wall air conditioner or heat pump system is installed in buildings requiring many temperature control zones such as (1)office buildings, (2)motels and hotels, (3)apartments and dormitories, (4)schools and other education buildings, and (5)areas of nursing homes or hospitals where air recirculation is allowed.

These units can be used for renovation of existing buildings, because existing heating systems can still be used. The equipment can be used in both low-and high-rise buildings. In buildings where a stack effect is present, use should be limited to those areas that have dependable ventilation and a tight wall of separation between the interior and exterior.

Part III HVAC systems and equipment

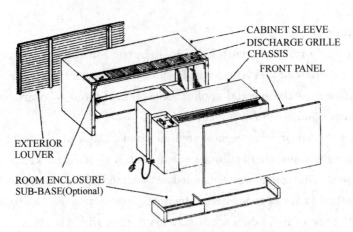

Fig. 3 Packaged Terminal Air Conditioner with
Combination Heating and Cooling Chassis

Room air conditioners are often used in parts of buildings primarily conditioned by other systems, especially where spaces to be conditioned are (1) physically isolated from the rest of the building and (2) occupied on a different time schedule (e.g., clergy offices in a church and ticket offices in a theater).

Ventilation air through each terminal may be inadequate in many situation, particularly in high-rise structures because of the stack effect. Electrically operated outdoor air dampers, which close automatically when the equipment is stopped, reduce heat losses in winter.

Refrigeration Equipment

Room air conditioners are generally supplied with hermetic reciprocating, or scroll compressors. Capillary tubes are used in place of expansion valves in most units.

Some room air conditioners have only one motor to drive both the evaporator and condenser fans. The unit circulates air through the condenser coil whenever the evaporator fan is running, even during the heating season. The annual energy consumption of a unit with a single motor is generally higher than one with a separate motor, even when energy efficiency ratio (coefficient of performance) is the same for both. The year-round continuous flow of air through the condenser increases dirt accumulation on the coil and other components, which increases maintenance costs and reduces equipment life.

Because through-the-wall conditioners are seldom installed with drains, they require a positive and reliable means of condensate disposal. Conditioners are available that spray the condensate in a fine mist over the condenser coil. These units dispose of more condensate than can be developed without any drip, splash, or spray. In heat pumps, provision must be made for disposal of condensate generated from the outside coil during the defrost cycle.

Many air-cooled room conditioners experience evaporator icing and become ineffective when the outdoor temperature falls below about 18℃. <u>Units that ice at a lower outdoor temperature may be required to handle the added load created by the high lighting levels and high solar radiation found in contemporary buildings.</u>

Heating Equipment

The air-to-air heat pump cycle is available in through-the-wall room air conditioners. Application considerations are quite similar to conventional units without the heat pump cycle, which is used for space heating when the outdoor temperature is above 2 to 5℃. Electric resistance elements supply heating below this level and during defrost cycles.

The prime advantage of the heat pump cycle is that it reduces the annual energy consumption for heating. Savings in heat energy over conventional electric heating ranges from 10 to 60%, depending on the climate.

Controls

All controls for through-the-wall air conditioners are included as a part of the conditioner. The following control configurations are available:

Thermostat Control. Thermostats are either unit mounted or remote wall mounted.

Guest Room Control for Motels and Hotels. This has provisions for starting and stopping the equipment from a central point.

Office Building and School Controls. These controls (for occupancies of less than 24h) start and stop the equipment at preset times with a time clock. The conditioners operate normally with the unit thermostat until the preset cutoff time. After this point, each conditioner has its own reset control, which allows the occupant of the conditioned space to reset the conditioner for either cooling or heating, as required.

Master/Slave Control. This type of control is used when multiple conditioners are operated by the same thermostat.

Emergency Standby Control. Standby control allows a conditioner to operate during an emergency, such as a power failure, so that the roomside blowers can operate to provide heating. Units must be specially wired to allow operation on emergency generator circuits.

Acoustics

The noise from these units may be objectionable and should be checked to ensure it meets sound level requirements.

INTERCONNECTED ROOM-BY-ROOM SYSTEMS

Multiple-unit systems generally use single-zone unitary air conditioners with a unit for each zone (Figure 4). Zoning is determined by (1) cooling and heating loads, (2) occupancy considerations, (3) flexibility requirements, (4) appearance considerations, and (5) equipment and duct space availability. Multiple-unit systems are popular for office buildings, manufacturing plants, shopping centers, department stores, and apartment buildings. Unitary self-contained units are excellent for renovation.

A common condensing and heat source loop connects all units together. Heat pumped into the loop by units in the cooling mode can be reclaimed by units in a heat pump heating mode. During moderate weather and in buildings with high diversity in cooling and heating loads, these systems will tend to balance the building load. Heat from warm areas are in effect moved to cooler areas of the building. This minimizes the amount of auxiliary loop heating and cooling required. Figure 5 shows a typical unit with some commonly used components.

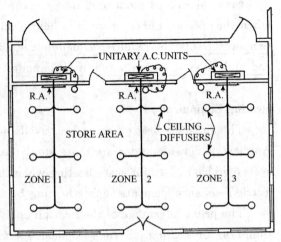

Fig. 4 Multiple Packaged Units

Advantages
- Installation is simple. Equipment is readily available in sizes that allow easy handling.
- Relocation of units to other spaces or buildings is practical, if necessary.
- Units are available with complete, self-contained control systems that include variable volume control, economizer cycle, night setback, and morning warm-up.
- Easy access to equipment facilitates routine maintenance.

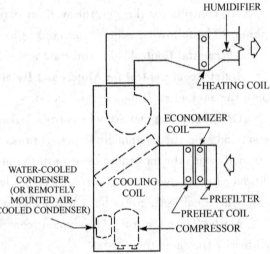

Fig. 5 Unitary Packaged Unit with Accessories

Disadvantages
- Fans may have limited static pressure ratings.
- Integral air-cooled units must be located along outside walls.
- Multiple units and equipment closets or rooms may occupy rentable floor space.
- Close proximity to building occupants may create noise problems.
- Discharge temperature may vary too much because of on-off or step control.

Design Considerations

Unitary systems can be used throughout a building or to supplement perimeter area packaged terminal units (Figure 6). Because core areas frequently have little or no heat loss, unitary equipment with water-cooled condensers can be applied with water-source heat pumps serving the perimeter.

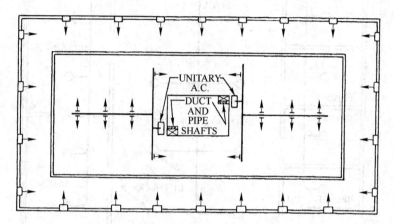

Fig. 6 Multiroom, Multistory Office Building with Unitary Core
and Through-the-Wall Perimeter Air Conditioners

In this multiple-unit system, one unit may be used to precondition outside air for a group of units (Figure 7). This all-outdoor air unit prevents hot, humid air from entering the conditioned space under periods of light load. The outdoor unit should have sufficient capacity to cool the required ventilation air from outdoor design conditions to interior design dew point. Zone units are then sized to handle only the internal load for their particular area.

Special-purpose unitary equipment is frequently used to cool, dehumidify, humidify, and reheat to maintain close control of space temperature and humidity in computer areas.

Refrigeration Equipment

Compressors are usually hermetic, reciprocating, or scroll compressors. Capillary tubes are used for expansion. Condensers are water cooled and connected to a central loop. A cooling tower or fluid cooler supplies supplemental cooling to the central loop as required.

Heating Equipment

The heating cycle on interconnected systems primarily uses an air-to-water heat pump. Supplemental heat to the loop is supplied with a central boiler.

Part III HVAC systems and equipment

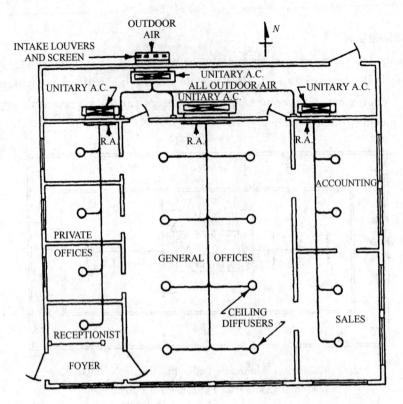

Fig. 7 Multiple-Packaged Units with Separate Outdoor Air Makeup Unit

Controls

Units under 70kW of cooling are typically constant-volume units. Variable-air-volume distribution is accomplished with a bypass damper that allows excess supply air to bypass to the return air duct. The bypass damper ensures constant airflow across the direct-expansion cooling coil to avoid coil freeze-up due to low air flow. The damper is usually controlled by the supply duct pressure.

Economizer Cycle.

When the outdoor temperature permits, energy use can be reduced in many locales by cooling with outdoor air in lieu of mechanical refrigeration. Units must be located close to an outside wall or outside air duct shafts. Where this is not possible, it may be practical to add an economizer cooling coil adjacent to the preheat coil (Figure 5). Cold water is obtained by cooling the condenser water through a winterized cooling tower.

Acoustics

Because these units are typically located near the occupied space, they can have a significant effect on acoustics. The designer must study both the airflow breakout path and the unit's radiated sound power when selecting wall and ceiling construction surrounding the unit. Locating units over non-critical work spaces such as restrooms or storage areas around the equipment room helps reduce noise in the occupied space.

WORDS AND EXPRESSIONS

acoustics	*n.* 声学
airflow breakout	气流进出口
airfoil	机翼型的
atomize	*v.* 使雾化，使成微粒
attenuate	*v.* 削弱
backward inclined	后倾的（后曲的）
cabinet	*n.* 机壳
capillary tube	毛细管
certified rating	经过鉴定的标定
chassis	*n.* 底盘
custom design	用户定制设计
custom program	自定义编程
de-energize	*v.* 不供电
defective	*adj.* 有缺陷的
deflection	*n.* 偏斜，偏转，偏差
direct-expansion cooling coil	直接蒸发式表面冷却器
discharge	*n.* 流出，排出
draw-through	抽出
duty cycle	工作循环
economizer	经济器（节能装置）
flash	*v.* 以薄的保护层覆盖
forward curved	前曲的
guide vane	导流叶片
hermetic	*adj.* 密封的
in lieu of	替代；顶替
inclement	*adj.* 险恶的，严酷的
interlock	*v.* 连锁
internally lined	内部加衬垫的
linkage	*n.* 连锁
louver	*n.* 百叶
malfunction	*n.* 故障
master/slave control	主从控制
neoprene pad	橡胶衬垫
octave band	倍频带
off-schedule	非工作日的

outweigh	v. 超过
packaged terminal air conditioner	组合式末端空调器
propane	n. ［化］丙烷
property line	地界
relief damper	排风阀
renovation	n. 翻修，修缮
routine unit maintenance	日常机组维护
scroll compressor	涡旋式压缩机
self-contained unit	独立式机组
service vestibule	维修前室
setback	延迟
silencer	n. 消声器
spare	n. 备件
spring isolator	弹簧隔振器
stand-alone	单机
supplement	v. 补充
surge region	喘振区
swing	n. 摇摆
unload	v. 卸载
vinyl	n. ［化］乙烯基

NOTATIONS

(1) Commercial grade unitary equipment packages are available only in pre-established increments of capacity with set performance parameters, such as the sensible heat ratio at a given room condition or the amount of airflow per kilowatt of refrigeration.

商业分级的单元式成套设备只在设定的性能参数，如给定房间状态的显热比或每千瓦制冷量的风量下，按预先制定的容量增量才可用。

(2) The wall box should drain to the outside.

暗线箱应对外排水。

(3) Because through-the-wall conditioners are seldom installed with drains, they require a positive and reliable means of condensate disposal. Conditioners are available that spray the condensate in a fine mist over the condenser coil. These units dispose of more condensate than can be developed without any drip, splash, or spray. In heat pumps, provision must be made for disposal of condensate generated from the outside coil during the defrost cycle.

因为穿墙式空调器很少装排水管，它们需要可靠的凝水处理(排出)。有些空调器将凝结水变成一种非常细小的水雾喷到冷凝器上，这些机组与其他不能滴、溅或喷凝结水的空调器相比能够排出更多的凝结水。对于热泵，必须保证室外盘管在除霜循环时产生的凝结水的排除。

(4) Units that ice at a lower outdoor temperature may be required to handle the added load created by the high lighting levels and high solar radiation found in contemporary buildings.

现代建筑的照明负荷和太阳辐射较大，使得负荷有所增加，可能需要在较低的室外温度下结冰的机组。

(5) Heat pumped into the loop by units in the cooling mode can be reclaimed by units in a heat pump heating mode.

处于供冷工况的机组排入环路中的热量可以被处于热泵供热工况的机组所回收。

EXERCISES

(1) 请翻译下列句子。

1) As airflow is decreased, the temperature of the air leaving the unit can often be reset upward so that a minimum ventilation rate can be maintained. Resetting the discharge air temperature is a way of demand limiting the unit, thus saving energy.

2) Because no energy is needed to transfer air or chilled water from mechanical equipment rooms, the energy consumption may be lower than for central systems. However, this advantage may be offset by the lower efficiency of this type of equipment.

3) In buildings where stack effect is present, use should be limited to those areas that have dependable ventilation and a tight wall of separation between the interior and exterior.

(2) 请用英语描述。

单元式空调器，分体式空调器，窗式空调器，组合式空调器

风冷式冷凝器，蒸发式冷凝器

离心式风机，轴流风机，前向叶型风机，后向叶型风机，机翼型风机

LESSON 10

AIR-COOLING AND DEHUMIDIFYING COILS

THE majority of the equipment used today for cooling and dehumidifying an airstream under forced convection incorporates a coil section that contains one or more cooling coils assembled in a coil bank arrangement. Such coil sections are used extensively as components in room terminal units; larger factory-assembled, self-contained air conditioners; central station air handlers; and field built-up systems. The applications of each type of coil are limited to the field within which the coil is rated. Other limitations are imposed by code requirements, proper choice of materials for the fluids used, the configuration of the air handler, and economic analysis of the possible alternatives for each installation.

USES FOR COILS

Coils are used for air cooling with or without accompanying dehumidification. Examples of cooling applications without dehumidification are (1) precooling coils that use well water or other relatively high-temperature water to reduce the load on the refrigerating equipment and (2) chilled water coils that remove sensible heat from chemical moisture-absorption apparatus. The heat-pipe coil is also used as a supplementary heat exchanger for preconditioning in airside sensible cooling. Most coil sections provide air sensible cooling and dehumidification simultaneously.

The assembly usually includes a means of cleaning air to protect the coil from accumulation of dirt and to keep dust and foreign matter out of the conditioned space. Although cooling and dehumidification are their principal functions, cooling coils can also be wetted with water or a hygroscopic liquid to aid in air cleaning, odor absorption, or frost prevention. Coils are also evaporatively cooled with a water spray to improve efficiency or capacity. For general comfort conditioning, cooling, and dehumidifying, the **extended surface (finned) cooling coil** design is the most popular and practical.

COIL CONSTRUCTION AND ARRANGEMENT

In finned coils, the external surface of the tubes is primary, and the fin surface is secondary. The primary surface generally consists of rows of round tubes or pipes that

---

Extracted from Chapter 21 of the ASHRAE *Handbook-HVAC Sysems and Equipment*.

may be staggered or placed in line with respect to the airflow. Flattened tubes or tubes with other nonround internal passageways are sometimes used. The inside surface of the tubes is usually smooth and plain, but some coil designs have various forms of internal fins or turbulence promoters(either fabricated or extruded)to enhance performance. The individual tube passes in a coil are usually interconnected by return bends(or hairpin bend tubes)to form the serpentine arrangement of multipass tube circuits. Coils are usually available with different circuit arrangements and combinations offering varying numbers of parallel water flow passes within the tube core (Figure 1).

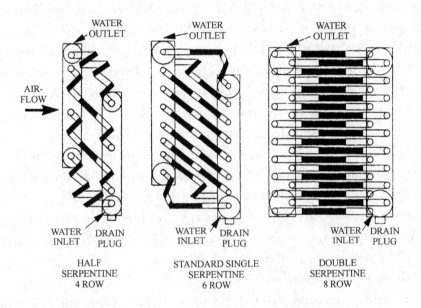

Fig. 1　Typical Water Circuit Arrangement

　　Cooling coils for water, aqueous glycol, brine, or halocarbon refrigerants usually have aluminum fins on copper tubes, although copper fins on copper tubes and aluminum fins on aluminum tubes(excluding water)are also used. Adhesives are sometimes used to bond header connections, return bends, and fin-tube joints, particularly for aluminum-to-aluminum joints. Certain special-application coils feature an all-aluminum extruded tube-and-fin surface.

　　Common core tube outside diameters are 8, 10, 12. 5, 16, 20, and 25mm, with fins spaced 1. 4 to 6. 4mm apart. Tube spacing ranges from 15 to 75mm on equilateral (staggered)or rectangular(in-line)centers, depending on the width of individual fins and on other performance considerations. Fins should be spaced according to the job to be performed, with special attention given to air friction; possibility of lint accumulation; and frost accumulation, especially at lower temperatures.

　　Tube wall thickness and the required use of alloys other than copper are determined mainly by the coil's working pressure and safety factor for hydrostatic burst(pressure).

Fin type and header construction also play a large part in this determination. Local job site codes and applicable nationally recognized safety standards should be consulted in the design and application of these coils.

Water and Aqueous Glycol Coils

Good performance of water-type coils requires both the elimination of all air and water traps within the water circuit and the proper distribution of water. Unless properly vented, air may accumulate in the coil tube circuits, reducing thermal performance and possibly causing noise or vibration in the piping system. Air vent and drain connections are usually provided on the coil water headers, but this does not eliminate the need to install, operate, and maintain the coil tube core in a level position. Individual coil vents and drain plugs are often incorporated on the headers (Figure 1). Water traps within the tubing of a properly leveled coil are usually caused by (1) improper nondraining circuit design and/or (2) center-of-coil downward sag. Such a situation may cause tube failure (e. g., freeze-up in cold climates or tube erosion due to untreated mineralized water).

Depending on performance requirements, the water velocity inside the tubes usually ranges from approximately 0.3 to 2.4m/s, and the design water pressure drop across the coils varies from about 15 to 150kPa. For nuclear HVAC applications, ASME *Standard* AG-1, Code on Nuclear Air and Gas Treatment, requires a minimum tube velocity of 0.6m/s. ARI *Standard* 410 requires a minimum of 0.3m/s or a Reynolds number of 3100 or greater. This yields more predictable performance.

In certain cases, the water may contain considerable sand and other foreign matter (e. g., in precooling coils using well water or in applications where minerals in the cooling water deposit on and foul the tube surface). It is best to filter out such sediment. Some coil manufacturers offer removable water header plates or a removable plug for each tube that allows the tube to be cleaned, ensuring a continuation of rated performance while the cooling units are in service. Where buildup of scale deposits or fouling of the water-side surface is expected, a scale factor is sometimes included when calculating thermal performance of the coils. Cupronickel, red brass, bronze, and other tube alloys help protect against corrosion and erosion deterioration caused primarily by internal fluid flow abrasive sediment. The core tubes of properly designed and installed coils should feature circuits that (1) have equally developed line length; (2) are self-draining by means of gravity during the coil's off cycle; (3) have the minimum pressure drop to aid in water distribution from the supply header without requiring an excessive pumping pressure; and (4) have equal feed and return by the supply and return header. Design for the proper in-tube water velocity determines the circuitry style required. Multirow coils are usually circuited to the cross-counterflow arrangement and oriented for top-outlet/bottom-feed connection.

Direct-Expansion Coils

Coils for halocarbon refrigerants present more complex cooling fluid distribution problems than do water or brine coils. The coil should cool effectively and uniformly throughout, with even refrigerant distribution. Halocarbon coils are used on two types of refrigerated systems: flooded and direct-expansion.

A flooded system is used mainly when a small temperature difference between the air and refrigerant is desired.

For direct-expansion systems, two of the most commonly used refrigerant liquid metering arrangements are the capillary tube assembly (or restrictor orifice) and the thermostatic expansion valve (TXV) device. The **capillary tube** is applied in factory-assembled, self-contained air conditioners up to approximately 35kW capacity but is most widely used on smaller capacity models such as window or room units. In this system, the bore and length of a capillary tube are sized so that at full load, under design conditions, just enough liquid refrigerant to be evaporated completely is metered from the condenser to the evaporator coil. While this type of metering arrangement does not operate over a wide range of conditions as efficiently as a thermostatic expansion valve system, its performance is targeted for a specific design condition.

A **thermostatic expansion valve** system is commonly used for all direct-expansion coil applications described in this chapter, particularly field-assembled coil sections and those used in central airhandling units and the larger, factory-assembled hermetic air conditioners. This system depends on the TXV to automatically regulate the rate of refrigerant liquid flow to the coil in direct proportion to the evaporation rate of refrigerant liquid in the coil, thereby maintaining optimum performance over a wide range of conditions. The superheat at the coil suction outlet is continually maintained within the usual predetermined limits of 3 to 6K. Because the TXV responds to the superheat at the coil outlet, the superheat within the coil is produced with the least possible sacrifice of active evaporating surface.

The length of each coil's refrigerant circuits, from the TXV's distributor feed tubes through the suction header, should be equal. The length of each circuit should be optimized to provide good heat transfer, good oil return, and a complementary pressure drop across the circuit. The coil should be installed level, and coil circuitry should be designed to self-drain by gravity toward the suction header connection.

To ensure reasonably uniform refrigerant distribution in multicircuit coils, a distributor is placed between the TXV and coil inlets to divide the refrigerant equally among the coil circuits. The refrigerant distributor must be effective in distributing both liquid and vapor because the refrigerant entering the coil is usually a mixture of the two, although mainly liquid by mass. Distributors can be placed in either the vertical or the horizontal position; however, the vertical down position usually distributes refrigerant between coil circuits better than the horizontal position for varying load conditions.

The individual coil circuit connections from the refrigerant distributor to the coil inlet are made of small-diameter tubing; the connections are all the same length and diameter so that the same flow occurs between each refrigerant distributor tube and each coil circuit. To approximate uniform refrigerant distribution, the refrigerant should flow to each refrigerant distributor circuit in proportion to the load on that coil. The heat load must be distributed equally to each of its refrigerant circuits to obtain optimum coil performance. If the coil load cannot be distributed uniformly, the coil should be recircuited and connected with more than one TXV to feed the circuits (individual suction may also help). In this way, the refrigerant distribution is reduced in proportion to the number of distributors that may have less of an effect on overall coil performance when the design must accommodate some unequal circuit loading. Unequal circuit loading may also be caused by such variables as uneven air velocity across the face of the coil, uneven entering air temperature, improper coil circuiting, oversized orifice in distributor, or the TXV's not being directly connected(close-coupled)to the distributor.

Control of Coils

Cooling capacity of water coils is controlled by varying either water flow or airflow. Water flow can be controlled by a three-way mixing, modulating, and/or throttling valve. For airflow control, face and bypass dampers are used. When cooling demand decreases, the coil face damper starts to close, and the bypass damper opens. In some cases, airflow is varied by controlling the fan capacity with speed controls, inlet vanes, or discharge dampers.

For factory-assembled, self-contained packaged systems or fieldassembled systems employing direct-expansion coils equipped with TXVs, a single valve is sometimes used for each coil; in other cases, two or more valves are used. The thermostatic expansion valve controls the refrigerant flow rate through the coil circuits so that the refrigerant vapor at the coil outlet is superheated properly. Superheat is obtained with suitable coil design and proper valve selection. Unlike water flow control valves, standard pressure/temperature-type thermostatic expansion valves alone do not control the refrigeration system's capacity or the temperature of the leaving air, nor do they maintain ambient conditions in specific spaces. However, some electronically controlled TXVs have these attributes.

In order to match the refrigeration load requirements for the conditioned space to the cooling capacity of the coil(s), a thermostat located in the conditioned space(s) or in the return air temporarily interrupts refrigerant flow to the direct-expansion cooling coils by stopping the compressor(s) and/or closing the solenoid liquid-line valve(s). Other solenoids unload compressors by means of suction control. For jobs with only a single zone of conditioned space, the compressor's on-off control is frequently used to modulate coil capacity.

Applications with multiple zones of conditioned space often use solenoid liquid-line valves to vary coil capacity. These valves should be used where thermostatic expansion valves feed certain types (or sections) of evaporator coils that may, according to load variations, require a temporary but positive interruption of refrigerant flow. This applies particularly to multiples of evaporator coils in a unit where one or more must be shut off temporarily to regulate its zone capacity. In such cases, a solenoid valve should be installed directly upstream of the thermostatic expansion valve(s). If more than one expansion valve feeds a particular zone coil, they may all be controlled by a single solenoid valve.

For a coil controlled by multiple refrigerant expansion valves, there are three arrangements: (1)face control, in which the coil is divided across its face; (2)row control; and (3)interlaced circuitry(Figure 2).

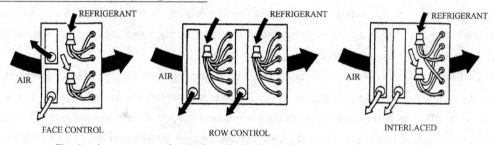

Fig. 2　Arrangements for Coils with Multiple Thermostatic Expansion Valves

Face control, which is the most widely used because of its simplicity, equally loads all refrigerant circuits within the coil. Face control has the disadvantage of permitting reevaporation of condensate on the coil portion not in operation and bypassing air into the conditioned space during partial load conditions, when some of the TXVs are on an off-cycle. However, while the bottom portion of the coil is cooling, some of the advantages of single-zone humidity control can be achieved with air bypasses through the inactive top portion.

Row control, seldom available as standard equipment, eliminates air bypassing during partial load operation and minimizes condensate reevaporation. Close attention is required for accurate calculation of row-depth capacity, circuit design, and TXV sizing.

Interlaced circuit control, uses whole face area and depth of coil when some of the expansion valves are shut off. Without a corresponding drop in airflow, modulating the refrigerant flow to an interlaced coil produces an increased coil surface temperature, thereby necessitating compressor protection(e. g., suction pressure regulators or compressor multiplexing).

Flow Arrangement

In the air-conditioning process, the relation of the fluid flow arrangement within the coil tubes to the coil depth greatly influences the performance of the heat transfer surface. Generally, air-cooling and dehumidifying coils are multirow and circuited for **counterflow** arrangement. The inlet air is applied at right angles to the coil's tube face (coil height), which is also at the coil's outlet header location. The air exits at the opposite face (side) of the coil where the corresponding inlet header is located. Counterflow can produce the highest possible heat exchange within the shortest possible (coil row) depth because it has the closest temperature relationships between tube fluid and air at each (air) side of the coil; the temperature of the entering air more closely approaches the temperature of the leaving fluid than the temperature of the leaving air approaches the temperature of the entry fluid. The potential of realizing the highest possible mean temperature difference is thus arranged for optimum performance.

Most direct-expansion coils also follow this general scheme of thermal counterflow, but the requirements for proper superheat control may necessitate a hybrid combination of parallel flow and counterflow. (Air flows in the same direction as the refrigerant in parallel flow operation.) Quite often, the optimum design for large coils is parallel flow arrangement in the coil's initial (entry) boiling region followed by counterflow in the superheat (exit) region. Such a hybrid arrangement is commonly used for process applications that require a low temperature difference (low TD).

Coil hand refers to either the right hand (RH) or left hand (LH) for counterflow arrangement of a multirow counterflow coil. There is no convention for what constitutes LH or RH, so manufacturers usually establish a convention for their own coils. Most manufacturers designate the location of the inlet water header or refrigerant distributor as the coil hand reference point. Figure 3 illustrates the more widely accepted coil hand designation for multirow water or refrigerant coils.

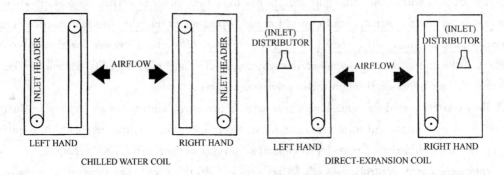

Fig. 3 Typical Coil Hand Designation

Applications

Figure 4 shows a typical arrangement of coils in a field built-up central station system. All air should be filtered to prevent dirt, insects, and foreign matter from accumulating on the coils. The cooling coil (and humidifier, when used) should include a drain pan under each coil to catch the condensate formed during the cooling cycle (and the excess water from the humidifier). The drain connection should be on the downstream side of the coils, be of ample size, have accessible cleanouts, and discharge to an indirect waste or

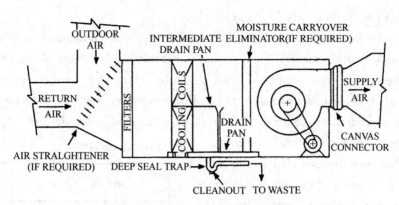

Fig. 4 Typical Arrangement of Cooling Coil Assembly in Built-Up
or Packaged Central Station Air Handler

storm sewer. The drain also requires a deep-seal trap so that no sewer gas can enter the system. Precautions must be taken if there is a possibility that the drain might freeze. The drain pan, unit casing, and water piping should be insulated to prevent sweating.

Factory-assembled central station air handlers incorporate most of the design features outlined for field built-up systems. These packaged units can generally accommodate various sizes, types, and row depths of cooling and heating coils to meet most job requirements. This usually eliminates the need for field built-up central systems, except on very large jobs.

The design features of the coil(fin spacing, tube spacing, face height, type of fins), together with the amount of moisture on the coil and the degree of surface cleanliness, determine the air velocity at which condensed moisture blows off the coil. Generally, condensate water begins to be blown off a plate fin coil face at air velocities above 3m/s. Water blowoff from the coils into air ductwork external to the air-conditioning unit should be prevented. However, water blowoff from the coils is not usually a problem if coil fin heights are limited to 1140mm and the unit is set up to catch and dispose of the condensate. When a number of coils are stacked one above another, the condensate is carried into the airstream as it drips from one coil to the next. A downstream eliminator section could prevent this, but an intermediate drain pan and/or condensate trough (Figure 5) to collect the condensate and conduct it directly to the main drain pan is preferred. Extending downstream of the coil, each drain pan length should be at least one-half the coil height, and somewhat greater when coil airflow

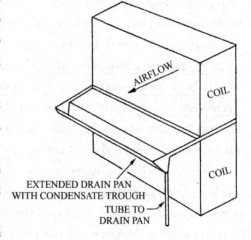

Fig. 5 Coil Bank Arrangement
with Intermediate Condensate Pan

face velocities and/or humidity levels are higher.

When water is likely to carry over from the air-conditioning unit into external air ductwork, and no other means of prevention is provided, eliminator plates should be installed on the downstream side of the coils. Usually, eliminator plates are not included in packaged units because other means of preventing carryover, such as space made available within the unit design for longer drain pan(s), are included in the design.

However, on sprayed-coil units, eliminators are usually included in the design. Such cooling and dehumidifying coils are sometimes sprayed with water to increase the rate of heat transfer, provide outlet air approaching saturation, and continually wash the surface of the coil. Coil sprays require a collecting tank, eliminators, and a recirculating pump (see Figure 6). Figure 6 also shows an air bypass, which helps a thermostat control maintain the humidity ratio by diverting a portion of the return air from the coil.

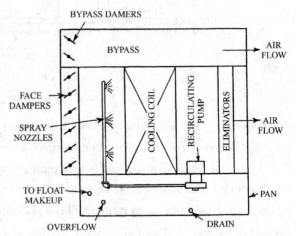

Fig. 6 Sprayed-Coil System with Air Bypass

In field-assembled systems or factory-assembled central station air-handling units, the fans are usually positioned downstream from the coil(s) in a draw-through arrangement. This arrangement provides acceptable airflow uniformity across the coil face more often than does the blow-through arrangement. In a blow-through arrangement, fan location upstream from the coils may require air baffles or diffuser plates between the fan discharge and the cooling coil to obtain uniform airflow. This is often the case in packaged multizone unit design. Airflow is considered to be uniform when the measured flow across the entire coil face varies no more than 20%.

<u>Air-cooling and dehumidifying coil frames, as well as all drain pans and troughs, should be of an acceptable corrosion-resistant material suitable for the system and its expected useful service life.</u> The air handler's coil section enclosure should be corrosion-resistant; be properly double-wall insulated; and have adequate access doors for changing air filters, cleaning coils, adjusting flow control valves, and maintaining motors.

Where suction line risers are used for air-cooling coils in direct-expansion refrigeration systems, the suction line must be sized properly to ensure oil return from coil to compressor at minimum load conditions. Oil return is normally intrinsic with factory-assembled, self-contained air conditioners but must be considered for factory-assembled central station units or field-installed cooling coil banks where suction line risers are required and are assembled at the job site.

LESSON 10

WORDS AND EXPRESSIONS

abrasive	*adj.*	使磨损的
aerodynamic	*adj.*	空气动力学的
air handler		空气处理装置（空调箱）
alloy	*n.*	合金
alternative refrigerant		替代制冷剂
aluminum	*n.*	铝
ample	*adj.*	充足的
aqueous glycol		乙二醇水溶液
baffle		导流片
barometric pressure		大气压力
blade	*n.*	叶片
blow-through		压入式
bore	*n.*	孔径
burr	*n.*	毛边
bypass damper		旁通风阀
carryover		遗留的水
catalog	*n.*	目录，目录册
chemical etching process		化学浸蚀过程
chemical moisture-absorption apparatus		化学吸湿装置
cleanliness	*n.*	清洁度
cleanout	*n.*	凝结水清除口（泄水孔）
compressor multiplexing		压缩机多级能量调节
computerized coil selection program		盘管选型计算机程序
constant pressure expansion valve		恒压膨胀阀
copper	*n.*	铜
corrosion	*n.*	侵蚀，腐蚀
cross-counterflow		交叉逆流
cupronickel	*n.*	白铜
deficiency	*n.*	缺乏，不足
dehumidify	*vt.*	除湿、减湿
deterioration	*n.*	劣化，老化
die-formed corrugation		凸凹波纹
drain plug		泄水阀
dual-path		双路的
durability	*n.*	经久，耐久力
eliminator plate		挡水板

eliminator section	挡水板段
equilateral	adj. 等边的
erosion	n. 腐蚀，侵蚀
ethylene glycol solution	乙二醇溶液
extruded	挤压而成的
flooded system	满液式系统（蒸发器）
hairpin bend tube	U 形弯头
halocarbon refrigerant	卤烃制冷剂
header connection	联箱
hybrid	adj. 混合的
hydrophilic	adj. ［化］亲水的，吸水的
impede	v. 妨碍，阻止
inadvertent	adj. 不注意的，疏忽的
inlet vane	进风导流叶片
interlaced	adj. 交织的，交错的
intrinsic	adj. 固有的
joint	n. 接缝，接合处，接合点
lanced fin	带毛刺式肋片
leading edge	最前端，前缘
lint	n. 细绒毛
louvered fin	带栅孔式肋片
low-side temperature glide	低端（蒸发）温度滑移
multipass	多通路的
psychrometrics	n. 焓湿图
rating	n. 评价，额定，标定
red brass	赤色黄铜
refrigerant blend	混合制冷剂
restrictor orifice	孔口节流阀
scale factor	水垢系数
sediment	n. 沉淀物
serpentine	adj. 蜿蜒的
solenoid valve	电磁阀
stagger	v. 交错放置
stock material	原料
suction line riser	吸气立管
suction pressure regulator	吸入压力调节阀
superheat	n. 过热度
surface tension	表面张力
temperature glide	温度滑移（非共沸混合制冷剂在定压下相变

LESSON 10

	（蒸发或凝结）时所伴随的温度变化
thermal resistance	热阻
thermostatic expansion valve（TXV）	热力膨胀阀
throttling valve	节流阀
trough	n. 槽，水槽
turbulence promoter	紊流强化装置
undulate	v. 使起伏，使波动
vent	n. 排气阀

NOTATIONS

(1) The majority of the equipment used today for cooling and dehumidifying an airstream under forced convection incorporates a coil section that contains one or more cooling coils assembled in a coil bank arrangement.

现在用于强制对流下冷却和减湿气流的设备大多数装有一个或多个冷却盘管按盘管组合布置组装的盘管段。

(2) The primary surface generally consists of rows of round tubes or pipes that may be staggered or placed in line with respect to the airflow.

主表面一般由与气流交错或成排的多排圆管构成。

(3) The inside surface of the tubes is usually smooth and plain, but some coil designs have various forms of internal fins or turbulence promoters (either fabricated or extruded) to enhance performance.

管的内表面通常是光滑和平坦的，但有些盘管设计具有不同形式的内肋或紊动强化装置（装配或挤压而成）以提高性能。

(4) Certain special-application coils feature an all-aluminum extruded tube-and-fin surface.

一些特殊用途的盘管以全铝挤压的肋管为特征。

(5) Tube wall thickness and the required use of alloys other than copper are determined mainly by the coil's working pressure and safety factor for hydrostatic burst (pressure).

管壁厚度和除铜以外需要使用合金主要由盘管的工作压力和静水压力的安全系数决定。

(6) Good performance of water-type coils requires both the elimination of all air and water traps within the water circuit and the proper distribution of water.

水类盘管的良好性能需要在水回路中消除所有空气和存水，并合理的配水。

(7) Water traps within the tubing of a properly leveled coil are usually caused by (1) improper nondraining circuit design and/or (2) center-of-coil downward sag.

一个合理找平的盘管的管内水的积存通常由两种原因引起：(1) 不当的无泄水回路设计和/或 (2) 盘管中心下垂。

(8) The core tubes of properly designed and installed coils should feature circuits that (1) have equally developed line length; (2) are self-draining by means of gravity during the

coil's off cycle; (3)have the minimum pressure drop to aid in water distribution from the supply header without requiring an excessive pumping pressure; and (4)have equal feed and return by the supply and return header.

合理设计和安装的盘管的芯管应有的特征(1)具有相等的展开管长;(2)在盘管不工作时能靠重力自行排水;(3)在从供水联箱布水时具有最小的压降,不要求过大的泵压;(4)通过送水联箱和回水联箱具有相同的供回水量。

(9) Multirow coils are usually circuited to the cross-counterflow arrangement and oriented for top-outlet/bottom-feed connection.

多排的盘管通常采用交叉逆流排列,并且取向为上出下进的联接。

(10) For direct-expansion systems, two of the most commonly used refrigerant liquid metering arrangements are the capillary tube assembly (or restrictor orifice) and the thermostatic expansion valve (TXV) device.

对于直接蒸发式系统,最常用的两种液态制冷剂配量装置是毛细管组件(或孔口节流阀)和热力膨胀阀(TXV)。

(11) The superheat at the coil suction outlet is continually maintained within the usual predetermined limits of 3 to 6K. Because the TXV responds to the superheat at the coil outlet, the superheat within the coil is produced with the least possible sacrifice of active evaporating surface.

在盘管吸气出口处的过热度要持续保持在通常预定的3到6K的限度内。因为热力膨胀阀受盘管出口的过热度控制,在盘管内的过热度是通过有效蒸发面积最少可能的损失产生的。

(12) For a coil controlled by multiple refrigerant expansion valves, there are three arrangements: (1)face control, in which the coil is divided across its face; (2)row control; and (3)interlaced circuitry.

由多个制冷剂膨胀阀控制的盘管,有三种配置:(1)迎面控制,盘管在迎风面上分开(并联);(2)排数控制(串联);(3)交错联接法。

(13) Face control has the disadvantage of permitting reevaporation of condensate on the coil portion not in operation and bypassing air into the conditioned space during partial load conditions, when some of the TXVs are on an off-cycle.

迎面控制具有在部分负荷工况使不运行盘管部分的凝结水再蒸发和旁通空气进入空调房间的缺点。

(14) Counterflow can produce the highest possible heat exchange within the shortest possible (coil row)depth because it has the closest temperature relationships between tube fluid and air at each(air)side of the coil; the temperature of the entering air more closely approaches the temperature of the leaving fluid than the temperature of the leaving air approaches the temperature of the entry fluid.

逆流能在最浅的可能深度(盘管排数)内产生最大可能的热交换,因为在盘管的每一侧逆流具有最接近的管内流体和空气的温度关系;进风的温度与液体的出口温度之差比出风温度与液体的进口温度之差还要小。

EXERCISES

(1) 请翻译下列句子。

1) Tube spacing ranges from 15 to 75mm on equilateral(staggered) or rectanguar(in-line) centers, depending on the width of individual fins and on other performance considerations.

2) Some coil manufacturers offer removable water header plates or a removable plug for each tube that allows the tube to be cleaned, ensuring a continuation of rated performance while the cooling units are in service.

3) If the coil load cannot be distributed uniformly, the coil should be recircuited and connected with more than one TXV to feed the circuits(individual suction may also help).

4) The design features of the coil (fin spacing, tube spacing, face height, type of fins), together with the amount of moisture on the coil and the degree of surface cleanliness, determine the air velocity at which condensed moisture blows off the coil.

5) Air-cooling and dehumidifying coil frames, as well as all drain pans and troughs, should be of an acceptable corrosion-resistant material suitable for the system and its expected useful service life.

(2) 请写出英文。

毛细管，热力膨胀阀，恒压阀，电磁阀，过热度

冷却盘管，挡水板，滴水盘，迎面风速，迎风面积，表面张力

LESSON 11

AIR CLEANERS FOR PARTICULATE CONTAMINANTS

THIS lesson discusses the cleaning of particulate contaminants from both ventilation air and recirculated air for the conditioning of building interiors. Complete air cleaning may also require the removal of airborne particles, microorganisms, and gaseous contaminants, but this chapter only covers the removal of airborne particles and briefly discusses bioaerosols.

The total suspended particulate concentration in the applications discussed in the chapter seldom exceeds $2mg/m^3$ and is usually less than $0.2mg/m^3$ of air. This contrasts with flue gas or exhaust gas from processes, where dust concentration typically ranges from 200 to $40000mg/m^3$.

With certain exceptions, the air cleaners addressed in this chapter are not used in exhaust gas streams, mainly because of the extreme dust concentration and temperature. However, the principles of air cleaning do apply to exhaust streams, and air cleaners discussed in the chapter are used extensively in supplying gases of low particulate concentration to industrial processes.

AIR CLEANING APPLICATIONS

Different fields of application require different degrees of air cleaning effectiveness. In industrial ventilation, removing only the larger dust particles from the airstream may be necessary for cleanliness of the structure, protection of mechanical equipment, and employee health. In other applications, surface discoloration must be prevented. Unfortunately, the smaller components of atmospheric dust are the worst offenders in smudging and discoloring building interiors. Electronic air cleaners or medium-to high-efficiency filters are required to remove smaller particles, especially the respirable fraction, which often must be controlled for health reasons. In cleanroom applications or when radioactive or other dangerous particles are present, high-or ultrahigh-efficiency filters should be selected.

Major factors influencing filter design and selection include (1) degree of air cleanliness required, (2) specific particle size range or aerosols that require filtration, (3) aerosol

---

Extracted from Chapter 24 of the ASHRAE *Handbook-HVAC Systems and Equipment*.

concentration, (4)resistance to airflow through the filter and (5)design face velocity to achieve published performance.

MECHANISMS OF PARTICLE COLLECTION

In the collection of particles, air cleaners made of fibrous media rely on the following five main principles or mechanisms:

Straining. The coarsest kind of filtration strains particles through an opening smaller than the particle being removed. It is most often observed as the collection of large particles and lint on the filter surface. The mechanism is not adequate to explain the filtration of submicrometre aerosols through fibrous matrices, which occurs through other physical mechanisms, as follows.

Inertial Impingement. When particles are large enough or of sufficient density that they cannot follow the air streamlines around a fiber, they cross over streamlines, impact on the fiber, and remain there if the attraction is strong enough. With flat panel and other minimal media area filters having high air velocities(where the effect of inertia is most pronounced), the particle may not adhere to the fiber because drag and bounce forces are so high. In this case, a viscous coating (preferably odorless and nonmigrating)is applied to the fiber to enhance retention of the particles. Such an adhesive coating is critical to the performance of metal mesh impingement filters.

Interception. Particles follow the air streamlines close enough to a fiber that the particle contacts the fiber and remains there mainly due to van der Waals forces. The process is dependent on the air velocity through the media being low enough not to dislodge the particles, and is therefore the predominate capture mechanism in extended media filters such as bag and deep-pleated rigid cartridge types.

Diffusion. The path of very small particles is not smooth but rather erratic and random within the air streamlines. This is caused by gas molecules in the air bombarding them(Brownian motion), producing an erratic path that brings the particles close enough to a media fiber to be captured by interception. As more and more particles are captured, a concentration gradient forms in the region of the fiber, further enhancing filtration by diffusion and interception. The effects of diffusion increase with decreasing particle size and media velocity.

Electrostatic Effects. Particle or media electrostatic charge can produce changes in the collection of dust affected by the electrical properties of the airstream. Some particles may carry a charge due to natural causes. Passive electrostatic(without a power source) filter fibers may be electrostatically charged during their manufacture or in some materials by predominately dry air blowing through the media. Charges on the particle and the media fibers can produce a strong attracting force if opposite. Efficiency is generally considered to be highest when the media is new and clean, decreasing rapidly as the filter loads.

Some progress has been made in calculating theoretical filter media efficiency from the physical constants of the media by considering the effects of the collection mechanisms (Lee and Liu 1982a, 1982b; Liu and Rubow 1986).

EVALUATING AIR CLEANERS

In addition to criteria affecting the degree of air cleanliness, factors such as cost(initial investment and maintenance), space requirements, and airflow resistance have encouraged the development of a wide variety of air cleaners. Accurate comparisons of different air cleaners can be made only from data obtained by standardized test methods.

The three operating characteristics that distinguish the various types of air cleaners are efficiency, resistance to airflow, and dust-holding capacity. **Efficiency** measures the ability of the air cleaner to remove particles from an airstream. Minimum efficiency during the life of the filter is the most meaningful characteristic for most filters and applications. **Resistance to airflow**(or simply resistance)is the static pressure drop differential across the filter at a given face velocity. The term static pressure differential is interchangeable with pressure drop and resistance if the difference of height in the filtering system is negligible. **Dust-holding capacity** defines the amount of a particular type of dust that an air cleaner can hold when it is operated at a specified airflow rate to some maximum resistance value (ASHRAE *Standard* 52.1).

Complete evaluation of air cleaners therefore requires data on efficiency, resistance, dust-holding capacity, and the effect of dust loading on efficiency and resistance. When applied to automatic renewable media devices(roll filters, for example), the evaluation must include the rate at which the media is supplied to maintain constant resistance when standardized test dust is fed at a specified rate.

Air filter testing is complex and no individual test adequately describes all filters. Ideally, performance testing of equipment should simulate the operation of the device under actual conditions and furnish an evaluation of the characteristics important to the equipment user. Wide variations in the amount and type of particles in the air being cleaned make evaluation difficult. Another complication is the difficulty of closely relating measurable performance to the specific requirements of users. Recirculated air tends to have a larger proportion of lint than does outside air. However, these difficulties should not obscure the objective that performance tests should strive to simulate actual use as closely as possible. In general, five types of tests, together with certain variations, determine air cleaner performance:

Arrestance. A standardized ASHRAE synthetic dust consisting of various particle sizes and types is fed into the test air stream to the air cleaner and the mass fraction of the dust removed is determined. In the ASHRAE *Standard* 52.1 test, summarized in the segment on Air Cleaner Test Methods in this chapter, this measurement is called **synthetic**

LESSON 11

dust arrestance to distinguish it from other efficiency values.

The indicated mass arrestance of air filters, as determined in the arrestance test, depends greatly on the particle size distribution of the test dust, which, in turn, is affected by its state of agglomeration. Therefore, this filter test requires a high degree of standardization of the test dust, the dust dispersion apparatus, and other elements of test equipment and procedures. This test is particularly suited to distinguish between the many types of low-to medium-efficiency air filters that are most commonly used on in recirculating systems with air handlers and fan coil units having minimal external static pressure capability. It does not adequately distinguish between filters of higher efficiency.

ASHRAE Atmospheric Dust-Spot Efficiency. Unconditioned atmospheric air is passed into the air cleaner under test and the discoloration level of the cleaned air(downstream of the test filter)on filter paper targets is compared with that of the unfiltered outside air(upstream of the test filter). The dust-spot test measures the ability of a filter to reduce the soiling of fabrics and building interior surfaces. Because these effects depend mostly on fine particles, this test is most useful for filters of higher efficiency. The variety and variability of atmospheric dust(McCrone et al. 1967, Whitby et al. 1958, Horvath 1967) may cause the same filter to test at different dust spot efficiencies at different locations(or even at the same location at different times). The accuracy of this test diminishes on filters of lower efficiency.

Fractional Efficiency or Penetration. Uniform-sized particles are fed into the air cleaner and the percentage removed by the cleaner is determined, typically by a photometer, optical particle counter, or condensation nuclei counter. In fractional efficiency tests, the use of uniform particle size aerosols has resulted in accurate measure of the particle size versus efficiency characteristic of filters over a wide atmospheric size spectrum. The method is timeconsuming and has been used primarily in research. However, the dioctyl phthalate(DOP)or Emery 3000 test for HEPA filters is widely used for production testing at a narrow particle size range. For more information on the DOP test, see the section on DOP Penetration Test.

Efficiency by Particle Size. A polydispersed challenge aerosol such as potassium chloride is metered into the test airstream to the air cleaner. Air samples taken upstream and downstream are drawn through an optical particle counter or similar measurement device to obtain removal efficiency versus particle size at a specific airflow rate.

Dust Holding Capacity. The true dust-holding capacity of similar air cleaners is a function of environmental conditions as well as the variability of atmospheric dust(size, shape and concentration)and is therefore impossible to duplicate in a laboratory test. For testing purposes, measured amounts of standardized dust is used to artificially load the filters. This procedure shortens the dust-loading cycle from weeks or years to hours. Artificial dusts are not the same as atmospheric dusts, so dust-holding capacity as measured by these accelerated tests are different from that achieved by "life" tests using atmospheric dust. The exact life of a filter in field use is impossible to determine by laboratory testing. However, testing of filters under standardized conditions does provide a rough guide to the relative effect of dust loading on the performance of similar units, and is one means used to compare them air cleaners.

Reputable laboratories perform accurate and reproducible filter tests within acceptable tolerances. Differences in reported values generally lie within the variability of the test aerosols, measurement devices, and dusts. Because most media are made of random air- or water-laid fibrous materials, the inherent media variations affect filter performance. Awareness of these variations prevents misunderstanding and specification of impossibly close performance tolerances. Caution must be exercised in interpreting published efficiency data, because the performance of two identical air cleaners tested by the same procedure may not result in exactly the same value, nor will the result necessarily be exactly duplicated in a subsequent test. Test values from different procedures generally cannot be compared. A performance test value of air cleaner efficiency is only a guide to the rate of soiling of a space or of mechanical equipment.

TYPES OF AIR CLEANERS

Common air cleaners are broadly grouped as follows:

Fibrous media unit filters, in which the accumulating dust load causes pressure drop to increse up to some maximum recommended value. During this period, efficiency normally increases. However, at high dust loads, dust may adhere poorly to filter fibers and efficiency drops due to off-loading. Filters in such condition should be replaced or reconditioned, as should filters that have reached their final (maximum recommended) pressure drop. This category includes viscous impingement and dry-type air filters, available in low-efficiency to ultrahigh-efficiency construction.

Renewable media filters, in which fresh media is introduced into the airstream as needed to maintain essentially constant resistance and, consequently, constant average efficiency.

Electronic air cleaners, which, if maintained properly by regular cleaning, have relatively constant pressure drop and efficiency.

Combination air cleaners, which combine the above types. For example, an electronic air cleaner may be used as an agglomerator with a fibrous media downstream to catch the agglomerated particles blown off the plates. Electrode assemblies have been installed in air-handling systems, making the filtration system more effective (Frey 1985,1986). Also, low efficiency pads, throw-away panels and automatically-renewable media roll filters, or low-to medium-efficiency pleated prefilters may be used upstream of a high-efficiency filter to extend the life of the better and more costly final filter. Charged media filters are also available that increase particle deposition on media fibers by an induced electrostatic field. With these filters, pressure loss increases as it does on a non-charged fibrous media filter. The benefits of combining different air cleaning processes vary. ASHRAE *Standard* 52.1 and 52.2 test methods may be used to compare the performance of combination air cleaners.

FILTER TYPES AND THEIR PERFORMANCE

Panel Filters

  Viscous Impingement Panel Filters are made up of coarse fibers with a high porosity. The filter media are generally coated with an odorless nonmigrating adhesive or other viscous substance, such as oil, which causes particles that impinge on the fibers to stick to them. Design air velocity through the media usually ranges from 1 to 4m/s. These filters are characterized by low pressure drop, low cost, and good efficiency on lint and larger particles (5μm and larger), but low efficiency on normal atmospheric dust. They are commonly made 13 to 100mm thick. Unit panels are available in standard and special sizes up to about 610mm by 610mm. This type of filter is commonly used in residential furnaces and air conditioning and is often used as a prefilter for higher-efficiency filters.

  A number of different materials are used as the filtering medium, including metallic wools, expanded metals and foils, crimped screens, random matted wire, coarse(15 to 60μm diameter)glass fibers, coated animal hair, vegetable fibers, synthetic fibers, and synthetic open-cell foams.

  Although viscous impingement filters usually operate in the range of 1.5 to 3m/s they may be operated at higher velocities. The limiting factor, other than increased flow resistance, is the danger of blowing off agglomerates of collected dust and the viscous coating on the filter.

  The loading rate of a filter depends on the type and concentration of the dirt in the air being handled and the operating cycle of the system. Manometers, static pressure differential gages, or pressure transducers are often installed to measure the pressure drop across the filter bank. From this measurement, it can be determined when the filter requires servicing. The final allowable pressure differential may vary from one installation to another; but, in general, viscous impingement filters are serviced when their operating resistance reaches 120Pa. Life Cycle Cost (LCC), including energy necessary to overcome the filter resistance, should be calculated in order to evaluate the overall cost of the filtering system. The decline in filter efficiency caused by dust coating the adhesive, rather than by the increased resistance because of dust load, may be the limiting factor in operating life.

  The manner of servicing unit filters depends on their construction and use. Disposable viscous impingement, panel-type filters are constructed of inexpensive materials and are discarded after one period of use. The cell sides of this design are usually a combination of cardboard and metal stiffeners. Permanent unit filters are generally constructed of metal to withstand repeated handling. Various cleaning methods have been recommended for permanent filters; the most widely used involves washing the filter with steam or water(frequently with detergent)and then recoating it with its recommended adhesive by dipping or spraying. Unit viscous filters are also sometimes arranged for in-place washing and recoating.

The adhesive used on a viscous impingement filter requires careful engineering. Filter efficiency and dust-holding capacity depend on the specific type and quantity of adhesive used; this information is an essential part of test data and filter specifications. Desirable adhesive characteristics, in addition to efficiency and dust-holding capacity, are (1)a low percentage of volatiles to prevent excessive evaporation; (2)a viscosity that varies only slightly within the service temperature range; (3)the ability to inhibit growth of bacteria and mold spores; (4)a high capillarity or the ability to wet and retain the dust particles; (5)a high flash point and fire point; and (6)freedom from odorants or irritants.

Dry-Type Extended-Surface Filters use media of random fiber mats or blankets of varying thicknesses, fiber sizes, and densities. Bonded glass fiber, cellulose fibers, wool felt, polymers, synthetics, and other materials have been used commercially. The media in filters of this class are frequently supported by a wire frame in the form of pockets, or V-shaped or radial pleats. In other designs, the media may be self-supporting because of inherent rigidity or because airflow inflates it into extended form such as with bag filters. Pleating of the media provides a high ratio of media area to face area, thus allowing reasonable pressure drop and low media velocities.

In some designs, the filter media is replaceable and is held in position in permanent wire baskets. In most designs, the entire cell is discarded after it has accumulated its maximum dust load.

The efficiency of dry-type air filters is usually higher than that of panel filters, and the variety of media available makes it possible to furnish almost any degree of cleaning efficiency desired. The dus-tholding capacities of modern dry-type filter media and filter configurations are generally higher than those of panel filters.

The placement of coarse prefilters ahead of extended-surface filters is sometimes justified economically by the longer life of the main filters. Economic considerations should include the prefilter material cost, changeout labor, and increased fan power. Generally, prefilters should be considered only if they can substantially reduce the part of the dust that may plug the protected filter. A prefilter usually has an arrestance of at least 70% but is commonly rated up to 92%. Temporary prefilters protecting higher efficiency filters are worthwhile during building construction to capture heavy loads of coarse dust. Filters of 95% DOP efficiency and greater should always be protected by prefilters of 80%~85% or greater ASHRAE average atmospheric dust-spot efficiency. A single filter gage may be installed when a panel prefilter is placed adjacent to a final filter. Because the prefilter is frequently changed on a schedule, the final filter pressure drop can be read without the prefilter in place every time the prefilter is changed. For maximum accuracy and economy of prefilter use, two gages can be used. Some air filter housings are available with pressure taps between the pre-and final filter tracks to accommodate this arrangement.

The initial resistance of an extended-surface filter varies with the choice of media and the filter geometry. Commercial designs typically have an initial resistance from 25 to 250Pa. It is customary to replace the media when the final resistance of 125Pa is reached for low resistance units and 500Pa for the highest resistance units. <u>Dry media providing higher orders of cleaning efficiency have a higher average resistance to airflow.</u> The operating resistance of the fully dust-loaded filter must be considered in the design, because that is the maximum resistance against which the fan operates. <u>Variable-air-volume and constant air volume system controls prevent abnormally high airflows or possible fan motor overloading from occurring when filters are clean.</u>

Flat panel filters with media velocity equal to duct velocity are made only with the lowest efficiency dry-type media(open cell foams and textile denier nonwoven media). Initial resistance of this group, at rated airflow, is generally between 10 and 60Pa. They are usually operated to a final resistance of 120 to 175Pa.

In extended-surface filters of the intermediate efficiency ranges, the filter media area is much greater than the face area of the filter; hence, velocity through the filter media is substantially lower than the velocity approaching the filter face. Media velocities range from 0.03 to 0.5m/s, although approach velocities run to 4m/s. Depth in direction of airflow varies from 50 to 900mm.

Filter media used in the intermediate efficiency range include (1)fine glass or synthetic fibers, 0.7 to 10μm in diameter, in mat form up to 13mm thick; (2)wet laid paper or thin nonwoven mats of fine glass fibers, cellulose, or cotton wadding; and (3)nonwoven mats of comparatively large diameter fibers(more than 30μm) in greater thicknesses(up to 50mm).

Electret filters are composed of electrostatically charged fibers. The charges on the fibers augment the collection of smaller particles by interception and diffusion(Brownian motion)with Coulomb forces caused by the charges on the fibers. There are three types of these filters: resin wool, electret, and an electrostatically sprayed polymer. The charge on the resin wool fibers is produced by friction during the carding process. <u>During production of the electret, a corona discharge injects positive charges on one side of a thin polypropylene film and negative charges on the other side. These thin sheets are then shredded into fibers of rectangular cross-section.</u> The third process spins a liquid polymer into fibers in the presence of a strong electric field, which produces the charge separation. The efficiency of the charged-fiber filters is due to both the normal collection mechanisms of a media filter(related to the fiber diameter)and the strong local electrostatic effects (related to the amount of electrostatic charge). The effects induce efficient preliminary loading of the filter to enhance the caking process. However, dust collected on the media can reduce the efficiency of electret filters.

Very high-efficiency dry filters, HEPA(high efficiency particulate air)filters, and ULPA(ultra low penetration air)filters are made in an extended-surface configuration of deep space folds of submicrometre glass fiber paper. Such filters operate at duct velocities near 1.3m/s, with resistance rising from 120 to more than 500Pa over their service life. These filters are the standard for cleanroom, nuclear and toxic-particulate applications.

Membrane filters are used predominately for air sampling and specialized small-scale applications where their particular characteristics compensate for their fragility, high resistance, and high cost. They are available in many pore diameters and resistances and in flat sheet and pleated forms.

Renewable media filters may be one of two types: (1)moving-curtain viscous impingement filters or (2)moving-curtain drymedia roll filter.

In one viscous type, random-fiber(nonwoven)media is furnished in roll form. Fresh media is fed manually or automatically across the face of the filter, while the dirty media is rewound onto a roll at the bottom. When the roll is exhausted, the tail of the media is wound onto the take-up roll, and the entire roll is thrown away. A new roll is then installed, and the cycle is repeated.

<u>Moving-curtain filters may have the media automatically advanced by motor drives on command from a pressure switch, timer, or media light-transmission control.</u> A pressure switch control measures the pressure drop across the media and switches on and off at chosen upper and lower set points. This control saves media, but only if the static pressure probes are located properly and unaffected by modulating outside air and return air dampers. Most pressure drop controls do not work well in practice. Timers and media light-transmission controls help to avoid these problems; their duty cycles can usually be adjusted to provide satisfactory operation with acceptable media consumption.

Filters of this replaceable roll design generally have a signal indicating when the roll of media is nearly exhausted. At the same time, the drive motor is de-energized so that the filter cannot run out of media. The normal service requirements involve insertion of a clean roll of media at the top of the filter and disposal of the loaded dirty roll. Automatic filters of this design are not, however, limited in application to the vertical position. Horizontal arrangements are available for use with makeup air units and air-conditioning units. Adhesives must have qualities similar to those for panel-type viscous impingement filters, and they must withstand media compression and endure long storage.

<u>The second type of automatic viscous impingement filter consists of linked metal mesh media panels installed on a traveling curtain that intermittently passes through an adhesive reservoir.</u> In the reservoir, the panels give up their dust load and, at the same time, take on a new coating of adhesive. The panels thus form a continuous curtain that moves up one face and down the other face. The media curtain, continually cleaned and renewed

with fresh adhesive, lasts the life of the filter mechanism. The precipitated captured dirt must be removed periodically from the adhesive reservoir. New installations of this type of filter are rare in North America, but are often found in Europe and Asia.

The resistance of both types of viscous impingement automatically renewable filters remains approximately constant as long as proper operation is maintained. A resistance of 100 to 125Pa at a face velocity of 2.5m/s is typical of this class.

Moving-curtain dry-media roll filters use random-fiber(nonwoven)dry media of relatively high porosity for general ventilation service. Operating duct velocities near 1m/s are generally lower than those of viscous impingement filters.

Special automatic dry filters are also available, which are designed for the removal of lint in textile mills, laundries and dry-cleaning establishments and for the collection of lint and ink mist in printing pressrooms. The medium used is extremely thin and serves only as a base for the buildup of lint, which then acts as a filter medium. The dirt-laden media is discarded when the supply roll is used up.

Another form of filter designed specifically for dry lint removal consists of a moving curtain of wire screen, which is vacuum cleaned automatically at a position out of the airstream. Recovery of the collected lint is sometimes possible with such a device.

ASHRAE arrestance, efficiency, and dust-holding capacities for typical viscous impingement and dry renewable media filters are listed in Table 1.

Performance of Renewable Media Filters(Steady-State Values)   Table 1

Description	Type of Media	ASHRAE Arrestance, %	ASHRAE Atmospheric Dust-Spot Efficiency, %	ASHRAE Dust-Holding Capacity, g/m^2	Approach Velocity, m/s
20 to 40μm glass and synthetic fibers, 50 to 65mm thick	Viscous impingement	70 to 82	<20	600 to 2000	2.5
Permanent metal media cells or overlapping elements	Viscous impingement	70 to 80	<20	NA (permanent media)	2.5
Coarse textile denier nonwoven mat, 12 to 25mm thick	Dry	60 to 80	<20	150 to 750	2.5
Fine textile denier nonwoven mat, 12 to 25mm thick	Dry	80 to 90	<20	100 to 550	1

Electronic Air Cleaners

Electronic air cleaners can be highly efficient filters using electrostatic precipitation to remove and collect particulate contaminants such as dust, smoke, and pollen. The designation electronic air cleaner denotes a precipitator for HVAC air filtration. The filter consists of an ionization section and a collecting plate section.

In the ionization section, small-diameter wires with a positive direct current potential between 6 and 25kV are suspended equidistant between grounded plates. The high voltage on the wires creates an ionizing field for charging particles. The positive ions created in the field flow across the airstream and strike and adhere to the particles, thus imparting a charge to them. The charged particles then pass into the collecting plate section.

The collecting plate section consists of a series of parallel plates equally spaced with a positive direct current voltage of 4 to 10kV applied to alternate plates. Plates that are not charged are at ground potential. As the particles pass into this section, they are attracted to the plates by the electric field on the charges they carry; thus, they are removed from the airstream and collected by the plates. Particle retention is a combination of electrical and intermolecular adhesion forces and may be augmented by special oils or adhesives on the plates. Figure 1 shows a typical electronic air cleaner cell.

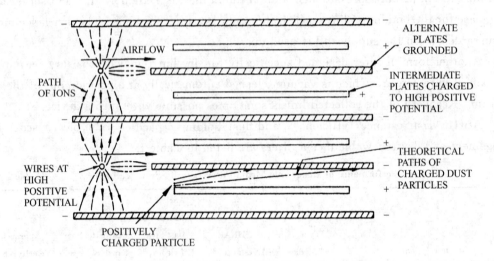

Fig. 1 Cross Section of Ionizing Electronic Air Cleaner

In lieu of positive direct current, a negative potential also functions on the same principle, but more ozone is generated. With voltages of 4 to 25kV(dc), safety measures are required. A typical arrangement makes the air cleaner inoperative when the doors are removed for cleaning the cells or servicing the power pack. Electronic air cleaners typically operate from a 120 or 240V(ac) single-phase electrical service. The high voltage supplied to the air cleaner cells is normally created with solid-state power supplies. The electrical power consumption ranges from 40 to 85W per m^3/s of air cleaner capacity.

This type of air filter can remove and collect airborne contaminants with an initial efficiency of up to 98% at low airflow velocities (0.8 to 1.8m/s) when tested according to ASHRAE *Standard* 52.1. Efficiency decreases (1) as the collecting plates become loaded with particulates, (2) with higher velocities, or (3) with nonuniform velocity.

As with most air filtration devices, the duct approaches to and from the air cleaner housing should be arranged so that the airflow is distributed uniformly over the face area.

Panel prefilters should also be used to help distribute the airflow and to trap large particles that might short out or cause excessive arcing within the high-voltage section of the air cleaner cell. Electronic air cleaner design parameters of air velocity, ionizer field strength, cell plate spacing, depth, and plate voltage must match the application requirements. These include contaminant type, particle size, volume of air, and required efficiency. Many units are designed for installation into central heating and cooling systems for total air filtration. Other self-contained units are furnished complete with air movers for source control of contaminants in specific applications that need an independent air cleaner.

Electronic air cleaner cells must be cleaned periodically with detergent and hot water. Some designs incorporate automatic wash systems that clean the cells in place; in other designs, the cells are removed for cleaning. The frequency of cleaning(washing)the cell depends on the contaminant and the concentration. Industrial applications may require cleaning every 8 hours, but a residential unit may only require cleaning at one to three month intervals. The timing of the cleaning schedule is important to keep the unit performing at peak efficiency. For some contaminants, special attention must be given to cleaning the ionizing wires.

Optional features are often available for electronic air cleaners. After-filters such as roll filters collect particulates that agglomerate and blow off the cell plates. These are used mainly where heavy contaminant loading occurs and extension of the cleaning cycle is desired. Cell collector plates may be coated with special oils, adhesives, or detergents to improve both particle retention and particle removal during cleaning. High-efficiency dry-type extended media area filters are also used as after-filters in special designs. The electronic air cleaner used in this system improves the service life of the dry filter and collects small particles such as smoke.

Another device, a **negative ionizer**, uses the principle of particle charging but does not use a collecting section. Particles enter the ionizer of the unit, receive an electrical charge, and then migrate to a grounded surface closest to the travel path.

Space Charge. Particulates that pass through an ionizer and are charged, but not removed, carry the electrical charge into the space. If continued on a large scale, a space charge builds up, which tends to drive these charged particles to walls and interior surfaces. Thus, a low-efficiency electronic air cleaner used in areas of high ambient dirt concentrations, or a malfunctioning unit, can blacken walls faster than if no cleaning device were used(Penney and Hewitt 1949, Sutton et al. 1964).

Ozone. All high-voltage devices are capable of producing ozone, which is toxic and damaging not only to human lungs, but to paper, rubber, and other materials. When properly designed and maintained, an electronic air cleaner produces an ozone concentration that only reach a fraction of the level acceptable for continuous human exposure and is less than that prevalent in many American cities (EPA 1996). Continuous arcing and brush discharge in an electronic air cleaner may yield an ozone level that is annoying or mildly toxic; this is indicated by a strong ozone odor. Although the nose is sensitive to ozone, only actual measurement of the concentration can determine that a hazardous condition exists.

ASHRAE *Standard* 62 defines acceptable concentrations of oxidants, of which ozone is a major contributor. The United States Environmental Protection Agency(EPA) specifies a 1-hour average maximum allowable exposure to ozone of 0.12ppm for outside ambient air. The United States Department of Health and Human Services specifies a maximum allowable continuous exposure to ozone of 0.05ppm for contaminants of indoor origin. Sutton et al. (1976) showed that indoor ozone levels are only 30% of the outdoor level with ionizing air cleaners operating, although Weschler et al. (1989) found that this level would increase when outdoor airflow is increased during an outdoor event that creates ozone.

WORDS AND EXPRESSIONS

agglomeration	*n.*	凝聚
arrestance	*n.*	捕集率
aspirator	*n.*	吸气器，吸出器
automatically-renewable media roll filter		自动更新介质的卷绕式过滤器
bombard	*v.*	碰撞，冲撞
caking	*n.*	烘烤，干燥
cardboard	*n.*	纸板
carding	*n.*	梳理（棉、毛、麻等）
cellulose fiber		纤维素纤维
charge separation		（等离子）电荷分离
consecutive	*adj.*	连续的
corona discharge		电晕放电
Coulomb force		库仑力
count median diameter		粒数中值径
cross-reference	*n.*	交叉参照
customary	*adj.*	习惯的，惯例的
designation	*n.*	指示，指定
diffusion		扩散效应
dioctyl phthalate (DOP)		邻苯二甲酸二辛酯
dislodge	*v.*	驱逐，移走
dust-spot efficiency		比色效率
dust-holding capacity		容尘量
electret	*n.*	驻极体

electronic air cleaner	静电空气洁净器
Electrostatic Effects	静电效应
emery	n. 金刚砂，刚玉砂
erratic	adj. 无确定路线，不稳定的
filter media(medium)	滤材
final filter	末级过滤器
fragility	n. 脆弱性
gasket	n. 垫圈，衬垫
impart	v. 给予
inertial impingement	惯性冲撞效应
inflate	v. 使膨胀，使充气
interception	拦截
manometer	n. 压力计
mass median diameter	质量中值径
media velocity	滤速
membrane filter	滤膜过滤器
opacity	n. 不透明度
patch	v. 修补，补缀
penetration	n. 穿透(率)
pinhole	n. 针孔，小孔
polydispersed aerosol	多分散气溶胶
polystyrene	n. 聚苯乙烯
porosity	n. 多孔性，孔隙率
potassium chloride	n. 氯化钾
power pack	电源组
precipitation	n. 沉降
pressure transducer	压力变送器
reproducible	adj. 可重复的
shock	n. 振动
smudge	v. 弄脏，染污
static pressure differential gage	静压差计
stiffener	n. 肋板
straining	n. 粗滤
submicrometre	n. 亚微米
underwriter	n. 保险商，担保人
van der Waals force	范德瓦尔斯力(原子与非极性分子之间微弱的吸引力)
viscous impingement air filter	黏性撞击式空气过滤器
wadding	n. 填料
wool felt	羊毛毯

NOTATIONS

(1) With flat panel and other minimal media area filters having high air velocities (where the effect of inertia is most pronounced), the particle may not adhere to the fiber because drag and bounce forces are so high. In this case, a viscous coating (preferably odorless and nonmigrating) is applied to the fiber to enhance retention of the particles.

用高速空气流的平板和其他最小介质面积的过滤器（惯性作用非常显著），微粒可能因为打滑和弹力很大而不会粘附到纤维上。在这种情况下，将一种黏性涂层加到过滤器上（比较好的是无味和非移动的）以促进粒子的滞留。

(2) When applied to automatic renewable media devices (roll filters, for example), the evaluation must include the rate at which the media is supplied to maintain constant resistance when standardized test dust is fed at a specified rate.

当应用于自动更新介质（滤材）的装置（例如，卷绕式过滤器）时，评价必须包括在用标准的测试粉尘以某一规定量送入时，保持阻力恒定所供给的介质（滤材）量。

(3) Ideally, performance testing of equipment should simulate the operation of the device under actual conditions and furnish an evaluation of the characteristics important to the equipment user.

理想地，设备的性能测试应该模拟在实际的运行条件下装置的运行，并且提供对设备用户重要的性能评价。

(4) This test is particularly suited to distinguish between the many types of low-to medium-efficiency air filters that are most commonly used on in recirculating systems with air handlers and fan coil units having minimal external static pressure capability.

这种测试特别适合于诸多低-中效空气过滤器之间加以区别，这些过滤器是普遍用于有最小外静压能力的带空气处理器和风机盘管的再循环系统。

(5) Charged media filters are also available that increase particle deposition on media fibers by an induced electrostatic field. With these filters, pressure loss increases as it does on a non-charged fibrous media filter.

带电介质过滤器也是可用的，它通过感应静电场增加粒子在介质纤维上的附着。就这些过滤器而言，压降（阻力）如同不带电的纤维介质过滤器一样增加。

(6) A number of different materials are used as the filtering medium, including metallic wools, expanded metals and foils, crimped screens, random matted wire, coarse (15 to 60 μm diameter) glass fibers, coated animal hair, vegetable fibers, synthetic fibers, and synthetic open-cell foams.

一些不同的材料用作过滤介质，包括金属棉、多孔金属网和薄片、波纹形筛网、任意纠结的金属丝、粗玻璃纤维（15~60μm 直径）、有涂层的动物毛发、植物纤维、合成纤维及合成开孔泡沫塑料。

(7) The cell sides of this design are usually a combination of cardboard and metal stiffeners.

这种设计的蜂窝侧通常是一种纸板和金属肋板的组合。

（8）Dry media providing higher orders of cleaning efficiency have a higher average resistance to airflow.

提供较高洁净效率级别的干式介质具有对气流较高的平均阻力。

（9）During production of the electret, a corona discharge injects positive charges on one side of a thin polypropylene film and negative charges on the other side. These thin sheets are then shredded into fibers of rectangular cross-section.

在驻极体生产期间，电晕放电在聚丙烯薄膜的一侧引入正电荷，另一侧引入负电荷。然后将这些薄膜切成矩形断面的纤维。

（10）Moving-curtain filters may have the media automatically advanced by motor drives on command from a pressure switch, timer, or media light-transmission control.

移动幕式过滤器可以按照压力开关、计时器或光透射控制的指令，由电机驱动介质自动前进。

（11）The second type of automatic viscous impingement filter consists of linked metal mesh media panels installed on a traveling curtain that intermittently passes through an adhesive reservoir.

第二种类型的自动黏性撞击型过滤器由安装在间歇地穿过一个粘合剂池的运动卷帘上的金属网板所构成。

（12）The filter consists of an ionization section and a collecting plate section.

过滤器由电离段和集尘板段组成。

（13）Particle retention is a combination of electrical and intermolecular adhesion forces and may be augmented by special oils or adhesives on the plates.

微粒的滞留是电和分子间粘合力的组合，还可能在集成板上涂特殊的油或粘结剂加以增强。

（14）请注意下列词语或短语的用法。

1) The path of very small particles is not(不是)smooth but rather(倒是)erratic and random within the air streamlines.

2) This measurement is called **synthetic dust arrestance** to distinguish(区别)it from other efficiency values.

3) In fractional efficiency tests, the use of uniform particle size aerosols has resulted in (导致)accurate measure of the particle size versus efficiency characteristic of filters over a wide atmospheric size spectrum.

4) The decline in filter efficiency caused by dust coating the adhesive, rather than(而不) by the increased resistance because of dust load, may be the limiting factor in operating life.

5) In lieu of(代替)positive direct current, a negative potential also functions on the same principle, but more ozone is generated.

6) Panel prefilters should also be used to help distribute the airflow and to trap large particles that might short out(短路)or cause excessive arcing within the high-voltage section of

the air cleaner cell.

EXERCISES

(1) 请解释下列句子中下划线词语的含义，并翻译整个句子。

1) As more and more particles are captured, a concentration gradient forms in the region of the fiber, further enhancing filtration by diffusion and interception.

2) Efficiency is generally considered to be highest when the media is new and clean, decreasing rapidly as the filter loads.

3) However, these difficulties should not obscure the objective that performance tests should strive to simulate actual use as closely as possible.

4) Air samples taken upstream and downstream are drawn through an optical particle counter or similar measurement device to obtain removal efficiency versus particle size at a specific airflow rate.

5) Variable air-volume and constant air volume system controls prevent abnormally high airflows or possible fan motor overloading from occurring when filters are clean.

6) In the ionization section, small-diameter wires with a positive direct current potential between 6 and 25kV are suspended equidistant between grounded plates.

(2) 请用英语表达。

粗效空气过滤器，中效空气过滤器，亚高效空气过滤器；

高效空气过滤器，超高效空气过滤器，黏性撞击式空气过滤器；

干式空气过滤器，静电空气洁净器，自动更新介质的卷绕式过滤器；

预过滤器，后置过滤器，末级过滤器；

主过滤器，带电介质空气过滤器，板式空气过滤器

LESSON 12

UNITARY AIR CONDITIONERS AND UNITARY HEAT PUMPS

THE AIR-CONDITIONING and Refrigeration Institute (ARI) defines a **unitary air conditioner** as one or more factory-made assemblies that normally include an evaporator or cooling coil and a compressor and condenser combination. It may include a heating function as well. ARI defines an **air-source unitary heat pump** as consisting of one or more factory-made assemblies, which normally include an indoor conditioning coil, compressor(s), and an outdoor coil. It must provide a heating function and possibly a cooling function as well. A **water-source heat pump** is a factory-made assembly that rejects or extracts heat to and from a water loop instead of from ambient air. A unitary air conditioner or heat pump having more than one factory-made assembly (e.g., indoor and outdoor units) is commonly called a **split system**.

Unitary equipment is divided into three general categories: residential, light commercial, and commercial. Residential equipment is single-phase unitary equipment with a cooling capacity of 19kW or less and is designed specifically for residential application. Light commercial equipment is generally three phase, with cooling capacity up to 40kW, and is designed for small businesses and commercial properties. Commercial unitary equipment has cooling capacity higher than 40kW and is designed for large commercial buildings.

In the development of unitary equipment, the following design objectives are considered: (1) user requirements, (2) application requirements, (3) installation, and (4) service.

User Requirements

The user primarily needs either space conditioning for comfort or a controlled environment for products or manufacturing processes. Cooling, dehumidification, filtration, and air circulation often meet those needs, although heating, humidification, and ventilation are also required in many applications.

Application Requirements

Unitary equipment is available in many secondary system configurations, such as

---

Extracted from Chapter 45 of the ASHRAE *Handbook-HVAC Systems and Equipment*.

- **Single zone, constant volume**, which consists of one controlled space with one thermostat that controls to maintain a set point.
- **Multizone, constant volume**, which has several controlled spaces served by one unit that supplies air of different temperatures to different zones as demanded (Figure 1).
- **Single zone, variable volume**, which consists of several controlled spaces served by one unit. Supply air from the unit is at a constant temperature, with air volume to each space varied to satisfy space demands (Figure 2).

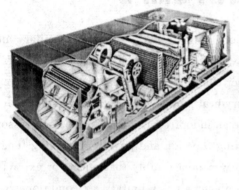

Fig. 1 Typical Rooftop Air-Cooled Single-Package Air Conditioner (Multizone)

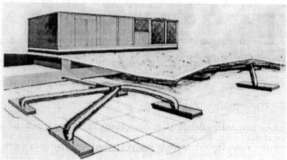

Fig. 2 Single-Package Air Equipment with Variable Air Volume

Such factors as size, shape, and use of the building; availability and cost of energy; building aesthetics (equipment located outdoors); and space available for equipment are considered to determine the type of unitary equipment best suited to a given application. In general, roof-mounted single-package unitary equipment is limited to five or six stories because duct space and available blower power become excessive in taller buildings. Indoor, single-zone equipment is generally less expensive to maintain and service than multizone units located outdoors.

The building load and airflow requirements determine equipment capacity, whereas the availability and cost of fuels determine the energy source. Control system requirements must be established, and any unusual operating conditions must be considered early in the planning stage. In some cases, custom-designed equipment may be necessary.

Manufacturers' literature has detailed information about geometry, performance, electrical characteristics, application, and operating limits. The system designer then focuses on selecting suitable equipment with the capacity for the application.

Installation

Unitary equipment is designed to keep installation costs low. The equipment must be installed properly so that it functions in accordance with the manufacturer's specifications. Interconnecting diagrams for the low-voltage control system should be documented for proper servicing in

LESSON 12

the future. Adequate planning for the installation of large, roof-mounted equipment is important because special rigging equipment is frequently required.

The refrigerant circuit must be clean, dry, and leak-free. An advantage of packaged unitary equipment is that proper installation minimizes the risk of field contamination of the circuit. Care must be taken to properly install split-system interconnecting tubing (e.g., proper cleanliness, brazing, and evacuation to remove moisture). <u>Some residential split systems are provided with precharged line sets and quick-connection couplings, which reduce the risk of field contamination of the refrigerant circuit.</u> Split systems should be charged according to the manufacturer's instructions.

In the installation of split systems, lines must be properly routed and sized to ensure good oil return to the compressor.

Unitary equipment must be located to avoid noise and vibration problems. Single-package equipment of over 70kW capacity should be mounted on concrete pads if vibration control is a concern. <u>Large-capacity equipment should be roof-mounted only after the structural adequacy of the roof has been evaluated.</u> If they are located over occupied space, roof-mounted units with return fans that use ceiling space for the return plenum should have a lined return plenum according to the manufacturer's recommendations. Duct silencers should be used where low sound levels are desired. Mass and sound data are available from many manufacturers. Additional installation guidelines include the following:

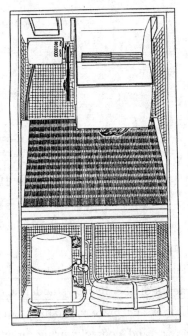

Fig. 3 Water-Cooled Single-Package Air Conditioner

- In general, install products containing compressors on solid, level surfaces.
- Avoid mounting products containing compressors (like remote units) on or touching the foundation of a house or building. A separate pad that does not touch the foundation is recommended to reduce any noise and vibration transmission through the slab.
- <u>Do not box in outdoor air-cooled units with fences, walls, overhangs, or bushes. Doing so reduces the air-moving capability of the unit, thus reducing efficiency.</u>
- For a split-system remote unit, choose an installation site that is close to the indoor portion of the system to minimize the pressure drop in the connecting tubing.
- Contact the unitary equipment manufacturer or consult the installation instructions for further information on installation procedures.

Unitary equipment should be listed or certified by nationally recognized testing laboratories to ensure safe operation and compliance with government and utility regulations.

part III *HVAC systems and equipment*

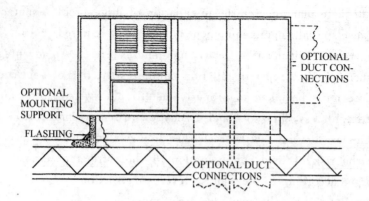

Fig. 4   Rooftop Installation of Air-Cooled Single-Package Unit

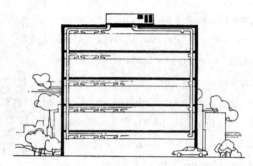

Fig. 5   Multistory Rooftop Installation
of Single-Package Unit

The equipment should also be installed to comply with the rating and application requirements of the agency standards to ensure that it performs according to industry criteria. Larger and more specialized equipment often does not carry agency labeling. However, power and control wiring practices should comply with the *National Electrical Code* (NFPA *Standard* 70). Local codes should be consulted before the installation is designed; local inspectors should be consulted before installation.

Service

A clear and accurate wiring diagram and a well-written service manual are essential to the installer and service personnel. Easy and safe service access must be provided in the equipment for periodic maintenance of filters and belts, cleaning, and lubrication. In addition, access for replacement of major components must be provided and preserved.

The availability of replacement parts aids proper service. Equitable warranty policies, covering 1 year of operation after installation, are offered by most manufacturers. Extended compressor warranties may be standard or optional.

Service personnel must be qualified to repair or replace mechanical and electrical com-

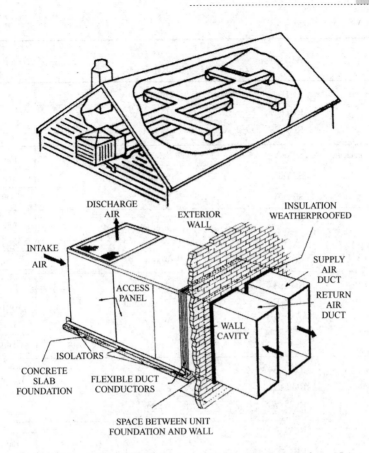

Fig. 6 Through-the-Wall Installation of Air-Cooled Single-Package Unit

ponents and to recover and properly recycle or dispose of any refrigerant removed from a system. They must also understand the importance of controlling moisture and other contaminants within the refrigerant circuit; they should know how to clean an hermetic system if it has been opened for service. Proper service procedures help ensure that the equipment will continue operating efficiently for its expected life.

TYPES OF UNITARY EQUIPMENT

Table 1 shows the types of unitary air conditioners available, and Table 2 shows the types of unitary heat pumps available. The following variations apply to some types and sizes of unitary equipment.

ARI Classification of Unitary Air Conditioners — Table 1

System Designation	ARI Type[a]	Heat Rejection	Arrangement	
Single package	SP-A	Air	Fan	Comp
	SP-E	Evap Cond	Evap	Cond
	SP-W	Water		

part III HVAC systems and equipment

continued

System Designation	ARI Type[a]	Heat Rejection	Arrangement
Refrigeration chassis	RCH-A	Air	Evap / Comp, Cond
	RCH-E	Evap Cond	
	RCH-W	Water	
Year-round single package	SPY-A	Air	Fan / Heat, Comp / Evap, Cond
	SPY-E	Evap Cond	
	SPY-W	Water	
Remote condenser	RC-A	Air	Fan / Evap / Comp ; Cond
	RC-E	Evap Cond	
	RC-W	Water	
Year-round remote condenser	RCY-A	Air	Fan / Evap / Heat / Comp ; Cond
	RCY-E	Evap Cond	
	RCY-W	Water	
Condensing unit, coil alone	RCU-A-C	Air	Evap / Comp ; Cond
	RCU-E-C	Evap Cond	
	RCU-W-C	Water	
Condensing unit, coil and blower	RCU-A-CB	Air	Fan / Evap ; Cond / Comp
	RCU-E-CB	Evap Cond	
	RCU-W-CB	Air	
Year-round condensing unit, coil and blower	RCUY-A-CB	Air	Fan / Evap / Heat ; Cond / Comp
	RCUY-E-CB	Evap Cond	
	RCUY-W-CB	Water	

[a] Adding a suffix of "—O" following any of the above classifications indicates equipment not intended for use with field-installed duct systems.

ARI Classification of Unitary Heat Pumps — Table 2

System Designation	ARI Type[a] Heating and Cooling	Heating Only	Arrangement
Single package	HSP-A	HOSP-A	Fan / Indoor Coil ; Comp / Outdoor Coil
	HSP-W	HOSP-W	
Remote outdoor coil	HRC-A-CB	HORC-A-CB	Fan / Indoor Coil / Comp ; Outdoor Coil
Remote outdoor coil with no indoor fan	HRC-A-C	HORC-A-C	Indoor Coil / Comp ; Outdoor Coil

continued

System Designation	ARI Type[a]		Arrangement	
	Heating and Cooling	Heating Only		
Split system	HRCU-A-CB	HORCU-A-CB	Fan | Indoor Coil	Comp | Outdoor Coil
	HRCU-W-CB	HORCU-W-CB		
Split system, no indoor fan	HRCU-A-C	HORCU-A-C	Indoor Coil	Comp | Outdoor Coil

[a] A suffix of "—O" following any of the above classifications indicates equipment not intended for use with field-installed duct systems.

Arrangement. Major unit components for various unitary air conditioners are arranged as shown in Table 1 and for unitary heat pumps as shown in Table 2.

Heat Rejection. Unitary air conditioner condensers may be air cooled, evaporatively cooled, or water cooled; the letters A, E, or W follow the ARI designation.

Heat Source/Sink. Unitary heat pump outdoor coils are designated as air-source or water-source by an A or W, following ARI practice. The same coils that act as a heat sink in the cooling mode act as the heat source in the heating mode.

Unit Exterior. The unit exterior should be decorative for inspace application, functional for equipment room and ducts, and weatherproofed for outdoors.

Placement. Unitary equipment can be mounted on floors, walls, ceilings, roofs, or a pad on the ground.

Indoor Air. Equipment with fans may have airflow arranged for vertical upflow or downflow, horizontal flow, 90° or 180° turns, or multizone. Indoor coils without fans are intended for forced-air furnaces or blower packages. Variable volume blowers may be incorporated with some systems.

Location. Unitary equipment intended for indoor use may be placed in the conditioned space with plenums or furred-in ducts or concealed in closets, attics, crawl spaces, basements, garages, utility rooms, or equipment rooms. Wall-mounted equipment may be attached to or built into a wall or transom. Outdoor equipment may be mounted on roofs or concrete pads on the ground. Installations must conform with local codes.

Heat. Unitary systems may incorporate gas-fired, oil-fired, electric, hot water coil, or steam coil heating sections. In unitary heat pumps, these heating sections supplement the heating capability.

Ventilation Air. Outdoor air dampers may be built into the equipment to provide outdoor air for cooling or ventilation.

Desuperheaters. Desuperheaters may be applied to unitary air conditioners and heat pumps. These devices recover heat from the compressor discharge gas and use it to heat domestic hot water. The desuperheater usually consists of a pump, a heat exchanger, and controls, and it can produce about 4.5mL/s of heated water per kilowatt of air conditioning (heating water from 15 to 54℃). Because desuperheaters improve cooling performance and reduce the degrading effect of cycling during heating, they are best applied where cooling requirements are high and where a significant number of heating hours occur above the building's balance point (Counts 1985). While properly applied desuperheaters can improve cooling efficiency, they can also reduce space-heating capacity. This causes the unit to run longer, which reduces the cycling of the system above the balance point.

Ductwork. Unitary eqipment is usually designed with fan capability for ductwork, although some units may be designed to discharge directly into the conditioned space.

Accessories. The manufacturer of any unitary equipment should be consulted before installing any accessories or equipment not specifically approved by the manufacturer. Such installations may not only void the warranty, but could cause the unitary equipment to function improperly or create fire or explosion hazards.

Combined Space-Conditioning/Water-Heating Systems

Unitary systems are available that provide both space conditioning and potable water heating. These systems are typically heat pumps, but some are available for cooling only. One type of combined system includes a full-condensing water-heating heat exchanger integrated into the refrigerant circuit of the space-conditioning system. Full-condensing system heat exchangers are larger than desuperheaters; they are generally sized to take the full condensing output of the compressor. Thus, they have much greater water-heating capacity. They also have controls that allow them to heat water year-round, either independently or coincidentally with space heating or space cooling. In spring and fall, the system is typically operated only to heat water.

Another type of combined system incorporates a separate, ancillary **heat pump water heater** (HPWH). The evaporator of this heater uses the return air (or liquid) stream of the space-conditoning system as a heat source. The HPWH thus cools the return stream during both space heating and cooling. In spring and fall, the space-conditioning blower (or pump) operates when water heating is needed.

As is the case with desuperheaters, simultaneous space and water heating reduces the output for space heating. This lower output is partially compensated for by the reduced cycling of the space-heating system above the balance point.

Combined systems can provide end users with significant energy savings and electric utilities with a significant reduction in demand. The overall performance of these systems is affected by the refrigerant charge and piping, the water piping, and the control logic and wiring. It is important, therefore, that the manufacturer's recommendations be closely followed. One special requirement is to locate the water-containing section(s) in areas not normally subjected to freezing temperatures.

Typical Unitary Equipment

Figures 1 through 6 show various types and installations of single-package equipment. Figure 7 shows a typical installation of a split-system, air-cooled condensing unit with indoor coil—the most widely used unitary cooling system. Figure 8 and Figure 9 also show split-system condensing units with coils and blower-coil units.

Many special light commercial and commercial unitary installations include a single-package air conditioner for use with variable air volume systems, as shown in Figure 2. These units are often equipped with a factory-installed system for controlling air volume in response to supply duct pressure (such as dampers or variable-speed drives).

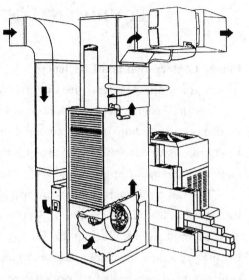

Fig. 7 Residential Installation of Split-System Air-Cooled Condensing Unit with Coil and Upflow Furnace

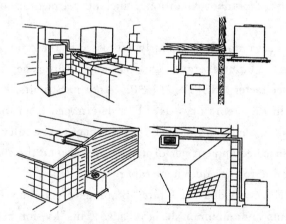

Fig. 8 Outdoor Installations of Split-System Air-Cooled Condensing Units with Coil and Upflow Furnace or with Indoor Blower-Coils

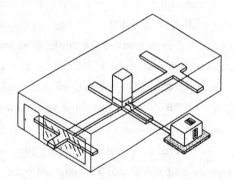

Fig. 9 Outdoor Installation of Split-System Air-Cooled Condensing Unit with Indoor Coil and Downflow Furnace

Another example of a specialized unit is the multizone unit shown in Figure 1. The manufacturer usually provides all controls, including zone dampers. The air path in these units is designed so that supply air may flow through a hot deck containing a means of heating or through a cold deck, which usually contains a direct-expansion evaporator coil.

To make multizone units more efficient, a control is commonly provided that locks out cooling by refrigeration when the heating unit is in operation and vice versa. Another variation to improve efficiency is the three-deck multizone. This unit has a hot deck, a cold deck, and a neutral deck carrying return air. Hot and/or cold deck air mixes only with air in the neutral deck.

EQUIPMENT AND SYSTEM STANDARDS

Energy Conservation and Efficiency

In the United States, the Energy Policy and Conservation Act, (Public Law 95-163) requires the Federal Trade Commission (FTC) to prescribe an energy label for many major appliances, including unitary air conditioners and heat pumps. The National Appliance Energy Conservation Act (NAECA) (Public Law 100-12) provides minimum efficiency standards for major appliances, including unitary air conditioners and heat pumps.

The U. S. Department of Energy (DOE) testing and rating procedure is documented in Appendix M to Subpart 430 of Section 10 of the *Code of Federal Regulations*, Uniform Test Method for Measuring the Energy Consumption of Central Air Conditioners. This testing procedure provides a seasonal measure of operating efficiency for residential unitary equipment. The **seasonal energy efficiency ratio** (SEER) is the ratio of the total seasonal cooling output to the total seasonal watt-hours of input energy. (SEER is not in SI units. To obtain COP, multiply SEER by 0.293.) This efficiency value is developed in the laboratory by conducting tests at various indoor and outdoor conditions, including a measure of performance under cyclic operation.

Seasonal heating mode efficiencies of heat pumps are similarly expressed as the ratio of the total heating output to the total seasonal input energy. This measure of efficiency is expressed as a **heating seasonal performance factor** (HSPF). (HSPF is not in SI units. To obtain COP, multiply HSPF by 0.293.) In the laboratory, HSPF is determined from the test results at different conditions, including a measure of cyclic performance. The calculated HSPF depends not only on the measured equipment performance, but also on the climatic conditions and the heating load relative to the equipment capacity.

For HSPF rating purposes, the DOE has divided the United States into six climatic regions and has defined a range of maximum and minimum design loads. This division has the effect of producing about 30 different HSPF ratings for a given piece of equipment. DOE has established Region 4 (moderate northern climate) and the minimum design load as being the typical climatic region and building design load to be used for comparative certified performance ratings.

SEER, HSPF, and operating costs vary appreciably with equipment design and size and from manufacturer to manufacturer. SEER and HSPF values, size ranges, and unit operating costs for DOE-covered unitary air conditioners that are certified by ARI are published semiannually in the ARI *Directory of Certified Unitary Products*.

In the United States, the Energy Policy Act of 1992 requires unitary equipment with cooling capacities from 19 to 70kW to meet the minimum efficiency levels prescribed by

LESSON 12

ASHRAE *Standard* 90.1.

ARI Certification Programs

Equipment up to 40kW. ARI conducts three certification programs relating to unitary equipment up to 40kW, which are covered in ARI *Standard* 210/240 and ARI *Standard* 270. These standards include the performance requirements necessary for good equipment design. They also include the methods of testing established by ASHRAE *Standard* 37.

As part of its certification program, ARI publishes the *Directory of Certified Unitary Products*. Issued twice a year, this directory identifies certified products enrolled in one or more programs and lists the certified capacity and energy efficiency for each unit. Certification involves the annual audit testing of approximately 30% of the basic unitary equipment models of each participating manufacturer.

ARI *Standard* 210/240 established definitions and classifications, testing and rating methods, and performance requirements. Ratings are determined at ARI standard rating conditions, rated nameplate voltage, and prescribed discharge duct static pressures with rated evaporator airflow not exceeding 60L/s per kilowatt. The standard requires that units dispose of condensate properly and have cabinets that do not sweat under cool, humid conditions. The ability to operate satisfactorily and restart at high ambient temperatures with low voltage is also tested.

For certification under ARI *Standard* 270, outdoor equipment is tested in accordance with Acoustical Society of America (ASA) *Standard* 92. Test results obtained on a one-third octave band basis are converted to a single number for application evaluation. Application principles are covered in ARI *Standard* 275.

Equipment over 40kW. Unitary air conditioners and heat pumps exceeding 40kW can be tested in accordance with ARI *Standard* 340/360. Unitary condensing units with capacities of 40kW or larger are covered by ARI *Standard* 365. ARI has a certification program for large unitary air-conditioning, heat pump, and condensing units with cooling capacities from 40 to 75kW. The ARI *Directory of Certified Applied Air-Conditioning Products*, published twice a year, contains the certified values for such equipment.

Safety Standards and Installation Codes

Approval agencies list unitary air conditioners complying with a standard like Underwriters Laboratories (UL) *Standard* 1995, Heating and Cooling Equipment (CSA *Standard* C22.2 No. 236). Other UL standards may also apply. An evaluation of the product determines that its design complies with the construction requirements specified in the standard and that the equipment can be installed in accordance with the applicable requirements of the *National Electrical Code*; ASHRAE *Standard* 15, Safety Code for Mechanical Refrigeration; NFPA *Standard* 90A, Installation of Air Conditioning and Ventilating Systems; and NFPA *Standard* 90B, Installation of Warm Air Heating and Air Condition-

part Ⅲ HVAC systems and equipment

ing Systems.

Tests determine that the equipment and all components will operate within their recognized ratings, including electrical, temperature, and pressure, when the equipment is energized at rated voltage and operated at specified environmental conditions. <u>Stipulated abnormal conditions are also imposed under which the product must perform in a safe manner.</u> The evaluation covers all operational features (such as electric space heating) that may be used in the product.

Products complying with the applicable requirements may bear the agency listing mark. An approval agency program includes the auditing of continued production at the manufacturer's factory.

WORDS AND EXPRESSIONS

accessories	*n.*	附件
accumulator	*n.*	贮液器
aesthetics	*n.*	美学
Air-conditioning and Refrigeration Institute (ARI)		美国空调与制冷学会
air-source unitary heat pump		空气源单元式热泵
appreciably	*adv.*	相当的，可观的
aquifer	*n.*	含水层，蓄水层
blance point		平衡点（热泵无需借助辅助热源就能满足供热需求的最低室外温度）
braze	*v.*	铜焊，钎焊
certification	*n.*	鉴定
desuperheater	*n.*	（过热蒸气的）冷却器
flooding	*n.*	溢流
furred-in		内部有衬垫的
line	*v.*	加内衬
lubrication	*n.*	润滑
propeller fan		轴流风机
pulley	*n.*	滑轮
refrigerant-reversing valve		制冷剂反向阀（四通阀）
slugging	*n.*	猛击
spike	*n.*	尖峰
split system		分体式系统
transom	*n.*	横梁
void	*v.*	使无效
warm-up	*n.*	预热

warranty	n. 质量保证，保修
water table	n. 地下水位
water-loop heat pump	水环热泵

NOTATIONS

(1) In general, roof-mounted single-package unitary equipment is limited to five or six stories because duct space and available blower power become excessive in taller buildings.

一般来说，屋顶单体单元式机组被限制在五层或六层的建筑使用，因为用于更高的建筑，管道空间和可用的风机功率变得过大。

(2) Some residential split systems are provided with precharged line sets and quick-connection couplings, which reduce the risk of field contamination of the refrigerant circuit.

有些住宅分体系统备有预充注的管线组和快速连接的接头，降低制冷剂回路现场污染的危险。

(3) Because desuperheaters improve cooling performance and reduce the degrading effect of cycling during heating, they are best applied where cooling requirements are high and where a significant number of heating hours occur above the building's balance point.

因为过热蒸汽冷却器能改善制冷性能，并能减小供热时循环的衰减效应，它们最好应用于高于建筑平衡点之上有很多供热小时数的场合。

(4) Full-condensing system heat exchangers are larger than desuperheaters; they are generally sized to take the full condensing output of the compressor.

全冷凝系统的热交换器比过热蒸汽冷却器大，它们通常以获取压缩机所释放的全部输出热量来定大小。

(5) DOE has established Region 4 (moderate northern climate) and the minimum design load as being the typical climatic region and building design load to be used for comparative certified performance ratings.

美国能源部已经建立第四气候带（中等北部气候）和最低设计负荷作为典型气候带和建筑设计负荷，用作比较鉴定合格的性能评价。

(6) Ratings are determined at ARI standard rating conditions, rated nameplate voltage, and prescribed discharge duct static pressures with rated evaporator airflow not exceeding 60L/s per kilowatt.

额定值是在 ARI 标准标定条件、额定名牌电压和规定的额定蒸发器风量不超过每千瓦 60L/s 时送风管静压下确定的。

(7) Stipulated abnormal conditions are also imposed under which the product must perform in a safe manner.

也要采用规定的非正常工况，在这些工况下产品机组必须以安全的状态运行。

(8) 请注意下列词语或短语的用法。

1) The equipment should also be installed to comply with(遵守)the rating and application

requirements of the agency standards to ensure that it performs according to industry criteria.

2) Service personnel must be qualified to repair or replace mechanical and electrical components and to recover and properly recycle or dispose of(处理,除去)any refrigerant removed from a system.

3) To make multizone units more efficient, a control is commonly provided that locks out (同步,锁定) cooling by refrigeration when the heating unit is in operation and vice versa.

EXERCISES

(1) 请解释下列句子中下划线词语的含义，并翻译整个句子。

1) Large-capacity equipment should be roof-mounted only after the structural adequacy of the roof has been evaluated.

2) Do not box in outdoor air-cooled units with fences, walls, overhangs, or bushes. Doing so reduces the air-moving capability of the unit, thus reducing efficiency.

3) The same coils that act as a heat sink in the cooling mode act as the heat source in the heating mode.

4) The air path in these units is designed so that supply air may flow through a hot deck containing a means of heating or through a cold deck, which usually contains a direct expansion evaporator coil.

5) The standard requires that units dispose of condensate properly and have cabinets that do not sweat under cool, humid conditions.

(2) 请用英语表达。

空气源热泵，水源热泵，水环热泵，地源热泵，单元式空调机组，分体式空调机组，季节能效比，制热季节性能系数

英文科研论文写作简介

1 引言

英文论文写作的前提是有创新研究成果，创新研究成果的关键是选题。科研论文写作常出现的一个误区是：以为好论文是"写"出来的，只要会写，论文总能被接受发表。其实，论文被发表只是结果，这个结果是和一系列科研环节密切相关的，论文写作只是其最后一个环节。在选择科研课题和工作切入点时，就需特别注意，一定要有创新内容，科学研究的灵魂是创新，重复别人的工作，从科研的角度来说，是没有意义的。值得注意的是，阅读有关英文科技论文，不仅可以了解研究进展和动态，而且，可以学会科技英文表达。同样，选题很好，研究工作做得不够细致、深入，也难有说服力，难以成为有价值的研究工作。由于本书只介绍英文科研论文的写作，不讲如何做研究，因此只介绍有了好的研究成果后如何写成合格的科研文章。

2 科研论文的一般格式

科研论文，不像散文、小说那样形式可以千姿百态，而是具有较为固定的格式。从某种意义上说，科研论文是"八股文"。

其通常的组成和每部分的特点可参见表1。

科研论文格式及其特点　　　　　　　　　　　　　　　　　　表1

组成部分名称（按文章顺序）	特点或简要说明
题目 Title	10～20 words 简明，不必求全。 Brief. A complete sentence is not necessary.
作者信息 　姓名 　单位地址 　联系方式：E-mail 地址、传真、电话 Authorship 　Names of authors 　Affiliation 　E-mail address and telephone and fax numbers for corresponding author, if possible	通讯作者：往往是固定研究人员或项目负责人。 Corresponding author: Faculty member or principal investigator.
摘要 Abstract	150～200 英文词，说明研究目的、方法、结果、结论和意义。可以写一些定量结果。不仅对读者，而且对文献检索者都有帮助。 150～200 words to give purpose, methods or procedures, new results and their significance, and conclusions. Write for literature searchers as well as Journal readers. Include major quantitative data if they can be stated briefly, but do not include background material.

续表

组成部分名称(按文章顺序)	特点或简要说明
关键词 Key words	3~5个关键词，作为论文检索用，使读者可用关键词方便的检索到此论文，并对论文按内容分类。 3~5 key words which can be regarded as index words.
符号表 Nomenclature，Notation or Symbols	说明文章中符号表示的量的意义，单位，尽量用国际单位制。 Use SI units as much as possible.
引言 Introduction	篇幅：全文的10%~20%。 说明所研究问题的重要性；相关研究回顾与综述；指出已有研究的不足和局限，但语气应友善而含蓄。说明本论文的目的和重要性。 Introduce the importance of the problem studied. Review of previous work. State the limitations or shortcomings of the previous work. Clearly state the purpose and significance of the present work. 注意：不必提及所有文献。如果一篇文献对所讨论专题做了综述，可只引用该文献，不必再重复标注所讨论的文献。一般情况下，引言不应超过3页(间行打印，无图表)。 Notice： Do not attempt to survey the literature completely. If a recent article has a survey on the subject, cite that article without repeating its individual citations. In general, the Introduction should be no more than 3 double-spaced word-processed pages with no figures and tables.
研究或实验方法 Research approach Theoretical section or Experimental section	篇幅：全文的20%~30%。 介绍为简化问题所作的必要且合理的假设； 对问题进行数学描述：列方程、边界条件和初始条件； 问题求解； 或介绍实验仪器、条件和步骤，使读者阅读后可重复实验。 Make necessary assumptions. Describe the problem in mathematical equations together with relating boundary and initial conditions. Obtain the solution. Let the research can be reproduced. ——Describe the apparatus and instruments. ——Describe pertinent and critical factors involved in the experimental work.
结果和讨论 Results and discussion	篇幅：全文的40%左右。 研究结果介绍，数据的必要解释，新发现的讨论，与其他相关结果的比较。 结果和讨论也可分开。 结果：直接的发现；讨论：间接的发现。 此部分内容安排要特别注意逻辑性。 Present the results. Discuss new findings. Provide explanations for data. Elucidate models. Compare the results with other related works. Results and Discussion may be separated. Results：direct findings. Discussion：indirect findings. Notice：please logically arrange the contents.
结论 Conclusions	介绍研究工作的主要结论。力求简明。 Summarize conclusions of the work.
致谢 Acknowledgement(s)	说明本工作受到的资助及得到的帮助。 Information regarding the supporter(s) (e. g., financial support) is included here.

续表

组成部分名称(按文章顺序)	特点或简要说明
参考文献 References	对于一般科研论文,参考文献为 10~20 篇;对于综述性论文,参考文献为 60~100 篇。 10~20 references for research paper and 60~100 references for review paper.
附录 Appendix	一些公式的详细推导等内容可放在附录部分,以便使论文更紧凑。 Some detailed derivation of equations etc. could be placed in this part.

以上为英文科技论文的一般要求,不同期刊的风格和要求会有所不同。

3 科技论文的写作步骤

步骤及注意事项如同绘画,绘画要构思、画轮廓、再描绘、修饰。科技论文的写作步骤可参见表 2。

英文科技论文写作步骤　　　　　　　　　　　　　　　　表 2

准备材料	
确定题目	
写提纲	和指导老师讨论
安排和调整材料	
写论文草稿	和指导老师讨论
认真检查:内容、炼字、炼句	
请指导教师修改	在有条件的情况下请 Native English speaker 修改英文

值得注意的是,论文最好在研究工作进行中就开始酝酿,这样对研究本身的完整性会有帮助,而且,写作过程中也往往会发现一些问题,由于实验装置尚在,实验还可进行,有些问题还可方便的解决。

4 各部分写作的注意事项

4.1 如何写英文摘要

英文摘要是全文的浓缩,一般包括研究目的、研究方法、研究结果和结论。摘要是全文的摘要,因此论文从引言(Introduction)、论文展开(Approach)、结果(Results)和讨论(Discussion)以及结论部分的要点在引言中都应有反映。摘要部分应尽可能简明,一般不超过 300 个词。摘要比论文具有更广泛的读者,因此,尽量用通俗易懂的词汇(这些词汇无需通过阅读全文或查阅相关文献后就可明白),且风格、时态等应统一。在英文摘要中,时态可以是一般现在时、一般过去时和现在完成时,具体用何种时态,应根据表达的内容而定,但一般多用被动语态。请看下面的例子。注意,摘要中别忘了写出论文的主要发现或结论。

一般情况下,摘要中不列参考文献,不含图表。英文摘要内容完整,可独立存在。摘要虽在最前面,但实际上,它往往最后写。等全文完成后,再根据全文的内容进行提炼和推敲。当然,有些国际会议,开始只需要提交摘要,这时,摘要常常先写。

下面列举了几篇国际期刊论文的英文摘要,供读者参考,同时请注意缩写字的使用。

例 1[1]

Abstract:Interactions between volatile organic compounds (VOCs) and vinyl flooring (VF), a relatively homogenous, diffusion-controlled building material, were characterized. The sorption/desorption behavior of VF was investigated using single-component and binary systems of seven common VOCs ranging in molecular weight from n-butanol to n-pentadecane. The simultaneous sorption of VOCs and water vapor by VF was also investigated. Rapid determination of the material/air partition coefficient (K) and the material-phase diffusion coefficient (D) for each VOC was achieved by placing thin VF slabs in a dynamic microbalance and subjecting them to controlled sorption/desorption cycles. K and D are shown to be independent of concentration for all of the VOCs and water vapor. For the four alkane VOCs studied, K correlates well with vapor pressure and D correlates well with molecular weight, providing a means to estimate these parameters for other alkane VOCs. While the simultaneous sorption of a binary mixture of VOCs is non-competitive, the presence of water vapor increases the uptake of VOCs by VF. This approach can be applied to other diffusion-controlled materials and should facilitate the prediction of their source/sink behavior using physically-based models.

Keywords:Building material; Emission; Indoor air; Microbalance; Sink; Sorption

例 2[2]

Abstract:Desiccant systems have been proposed as energy saving alternatives to vapor compression air conditioning for handling the latent load. Use of liquid desiccants offers several design and performance advantages over solid desiccants, especially when solar energy is used for regeneration. For liquid-gas contact, packed towers with low pressure drop provide good heat and mass transfer characteristics for compact designs. This paper presents the results from a study of the performance of a packed tower absorber and regenerator for an aqueous lithium chloride desiccant dehumidigication system. The rates of dehumidification and regeneration, as well as the effectiveness of the dehumidification and regeneration processes were assessed under the effects of variables such as air and desiccant flow rates, air temperature and humidity, and desiccant temperature and concentration. A variation of the öberg and Goswami Mathematical model was used to predict the experimental findings giving satisfactory results.

例 3[3]

Abstract:This paper presents a performance evaluation of two passive cooling strategies, daytime ventilation and night cooling, for a generic, six-story suburban apartment building in Beijing and Shanghai. The investigation uses a coupled, transient simulation approach to model heat transfer and airflow in the apartments. Wind-driven ventilation is simulated using computational fluid dynamics (CFD). Occupant thermal comfort is accessed using Fanger's comfort model. The results show that night cooling is superior to daytime ventilation. Night cooling may replace air-conditioning systems for a significant part of the cooling season in Beijing, but with a high condensation risk. For Shanghai, neither of the two passive cooling strategies can be considered successful.

4.2 如何写引言

中国有句俗话：好的开头等于成功的一半。英文中有句名言："A bad beginning makes a bad ending"。两者表达方式不同，意思却相近，开头对很多事非常重要，对写文章也不例外。引言即是文章的开头。

写作之前，心中需对阅读对象有所了解和估计，这样在行文时对遣词造句就会有把握，既避免过于专业，使读者难以理解，又不致过于平白，让读者的阅读索然无味。

引言一般用一般现在时写，如前所述，引言中应介绍以下几方面的内容：

(1) 介绍讨论的问题、研究的背景，说明讨论的范围及解决问题的重要性。读者往往通过浏览论文题目、摘要、引言、图标和结论决定是否仔细阅读全文。因此，在引言中应开门见山，说明要讨论的问题及其重要性。

(2) 相关研究回顾与综述。对已有研究的评价要实事求是，对前人工作的精彩和可参考之处应简要说明，对已有研究的不足和局限也应指出，但语气应友善而含蓄。

(3) 说明本研究的目的和特别之处。有了前面两部分的铺垫，现在就要具体说明本研究要解决什么问题，在解决思路、方法、手段上有什么新颖或改进之处。

(4) 介绍一下文章的结构安排。这是全部论文的导读，就像领人去一个地方游览或参观，先介绍一下游览的活动安排并给他一张游览地图。在这部分，下面的表述可供参考：

This paper is divided into five major sections as follows…

Section one of this paper opens with…

Section three develops the second hypothesis on…

Section four shows (introduces, reveals, treats, develops, deals with, etc.)…

The result of … is given in the last section.

(5) 介绍一下主要结论。

(4)和(5)的安排比较灵活，有时可不同时出现，甚至都不出现，只介绍(3)：本论文的目的或主要贡献及其重要性即可。

关于引言的功能，Raleigh Nelson 有一段形象的介绍："(it) may be thought of as a preliminary conference in which the writer and prospective reader 'go into a huddle' and agree in advance on the exact limits of the subject, the terms in which to discuss it, the angle from which to approach it, and the plan of treatment that will be most convenient to both."

引言部分逻辑性很强。首先当然是点出问题，并使读者一下子被吸引。这就必须交代为什么你选择该问题，该问题的解决状况如何，还有哪些问题需要研究，你如何解决这些问题，得到了哪些有意义的结果。这些环节联系紧密、环环相扣。

引言中要引用已发表的相关文献，一般有两种引出方式：按所引文献出现的先后顺序标注；按所引文献作者姓名的字母顺序标注。具体方式，视所投期刊要求而定。

下面通过一些例子对上面的介绍加以说明。

例 1[1]

INTRODUCTION

A variety of building materials (e. g., adhesives, sealants, paints, stains, carpets, vinyl flooring, and engineered woods) can act as indoor sources of volatile organic compounds (VOCs). Following their installation or application, these materials typically contain residual quantities of VOCs that are then emitted over time. Once installed and depending upon their properties, these materials may also interact with airborne VOCs through alternating sorption and desorption cycles (Zhao et al., 1999b, 2001). Consequently, building materials can have a significant impact on indoor air quality both as sources of and sinks for volatile compounds.

Current methods for characterizing the source/sink behavior of building materials typically involve chamber studies. This approach can b time-consuming and costly, and is subject to several limitations (Little and Hodgson, 1996). For those indoor sources and sinks that are controlled by internal diffusion processes, physically-based diffusion models hold considerable promise for prediction emission characteristics when compared to empirical methods (Cox et al., 2000b, 2001b).

The key parameters for physically-based models are the material/air partition coefficient (K), the material-phase diffusion coefficient (D), and, in the case of a source, the initial concentration of VOC in the material (C_0). Rapid and reliable determination of these key parameters by direct measurements or by estimations based on readily available VOC/building material properties should greatly facilitate the development and use of mechanistic models for characterizing the source/sink behavior of diffusion-controlled materials (Zhao et al., 1999a; Cox et al., 2000a, 2001a).

Several procedures have been used to measure D and K of volatile compounds in building materials. D and K have been inferred from experimental data obtained in chamber studies (Little et al., 1994). A procedure using a two-compartment chamber has also been used for D and K measurement. A specimen of building material is installed between the two compartments. A concentration of a particular compound if introduced into the gas-phase of one compartment while the gas-phase concentration in the other compartment is measured over time. D and K are then indirectly estimated from gas-phase concentration data (Bodalal et al., 2000; Meininghaus et al., 2000). A complicating feature of this method is that VOC transport between chambers may occur by gas-phase diffusion through pores in the building material in addition to solid-phase Fickian diffusion, confounding estimates of the mass transport characteristics of the solid material.

A procedure based on a European Committee for Standardization (CEN) method has also been used to estimate D. A building material sample is tightly fastened to the open end of a cup containing a liquid VOC. As the VOC diffuses from the saturated gas-phase through the building material sample, cup weight over time is recorded. Weight change data can be used to estimate D. (kirchner et al., 1999). A significant drawback of this method is that D has been shown to become concentration dependent in polymers at concentrations approaching saturation (Park et al., 1989).

In accordance with a previously proposed strategy for characterizing homogeneous, diffusion-controlled, indoor sources and sinks (Little and Hodgson, 1996), the objectives of this study were to (1) develop a simple and rapid experimental method for directly measuring the key equilibrium and kinetic parameters, (2) examine the validity of several primary assumptions upon which the previously mentioned physically-based models are founded and (3) develop correlations between the O and K, and readily available properties of VOCs.

例 2[2]

INTRODUCTION

Liquid desiccant cooling systems have been proposed as alternatives to the conventional vapor compression cooling systems to control air humidity, especially in hot and humid areas. Research has shown that a liquid desiccant cooling system can reduce the overall energy consumption, as well as shift the energy use away from electricity and toward renewable and cheaper fuels (Oberg and Goswami, 1998a). Burns et al. (1985) found that utilizing desiccant cooling in a supermarket reduced the energy cost of air conditioning by 60% as compared to conventional cooling. Oberg and Goswami (1998a) modeled a hybrid solar cooling system obtaining an electrical energy savings of 80%, and Chengchao and Ketao (1997) showed by computer simulation that solar liquid desiccant air conditioning has advantages over vapor compression air conditioning system in its suitability for hot and humid areas and high air flow rates.

Use of liquid desiccants offers several design and performance advantages over solid desiccants, especially when solar energy is used for regeneration (Oberg and Goswami, 1998c). Several liquid desiccants are commercially available: triethylene glycol, diethylene glycol, ethylene glycol, and brines such as calcium chloride, lithium chloride, lithium bromide, and calcium bromide which are used singly or in combination. The usefulness of a particular liquid desiccant depends upon the application. At the University of Florida, Oberg and Goswami (1998a, b) conducted a study of a hybrid solar liquid desiccant cooling system using triethylene glycol (TEG) as the desiccant. Their experimental work concluded that glycol works well as a desiccant. However, pure triethylene glycol does have a small vapor pressure which causes some of the glycol to evaporate into the air. Although triethylene glycol in nontoxic, any evaporation into the supply air stream makes it unacceptable for use in air conditioning of an occupied building. Therefore, there is a need to evaluate other liquid desiccants for hybrid solar desiccant cooling systems. Lithium chloride (LiCl) is a good candidate material since it has good desiccant characteristics and does not vaporize in air at ambient conditions. A disadvantage with LiCl is that it is corrosive. This paper presents an experimental and theoretical study of aqueous lithium chloride as a desiccant for a solar hybrid cooling system, using a packed bed dehumidifier and regenerator.

A number of experimental studies have been carried out on packed bed dehumidifiers using salt solutions as desiccants. Chung et al. (1992, 1993), and Chen et al. (1989) used lithium chloride (LiCl); Ullah et al. (1998), Kinsara et al. (1998) and Lazzarin et al. (1999) used calcium chloride (CaCl2); while Ahmed et al. (1997) and Patnaik et al. (1990) used lithium bromide (LiBr). Other experiments for absorbers using LiCl were carried out by Kessling et al. (1998), Kim et al. (1997) and Scalabrin and Scaltriti (1990).

The moisture that transfers from the air to the liquid desiccant in the dehumidifier causes a dilution of the desiccant resulting in a reduction in its ability to absorb more water. Therefore, the desiccant must be regenerated to its original concentration. The regeneration process requires heat which can be obtained from a low temperature source, for which solar energy and waste energy from other processes are suitable. Different ways to regenerate liquid desiccants have been proposed. Hollands (1963) presented results from the regeneration of lithium chloride in a solar still. Hollands focused his study on the still efficiency, concluding that lithium chloride can be regenerated in a solar still with a daily efficiency of 5 to 20% depending on the insolation and the concentration of the desiccant. Ahmed Khalid et al. (1998) presented an exergy analysis of a partly closed solar generator to compare it with the solar collector reported previously. Ahmed et al. (1997) simulated a hybrid cycle with a partly closed-open solar regenerator for regeneration the weak solution. They found that the system COP is about 50% higher than that of a conventional vapor absorption machine. Leboeuf and Lof (1980) presented an analysis of a lithium chloride open cycle absorption air conditioner which utilizes a packed bed for regeneration of the desiccant solution driven by solar heated air. In this case, the air temperature ranged from 65 to 96℃ while the desiccant temperature ranged from 40 to 55℃. Lof et al. (1984) conducted experimental and theoretical studies of regeneration of aqueous lithium chloride solution with solar heated air in a packed column. In this case, air at a temperature of 82 to 109℃ was used to regenerate the desiccant at an average temperature of 36℃.

In any thermodynamic system, the conditions of the working fluids and parameters of the physical equipment define the overall performance of the system. In a liquid desiccant cooling system, variables such as air and desiccant flow rate, air temperature and humidity, desiccant temperature and concentration are of great interest for the performance of the dehumidifier. The mass ratio of air to desiccant solution $MR = m_{air}/m_{sol}$ is an important factor for absorber efficiency and system capacity. Previous studies have reported the performance of packed bed absorbers and regenerators with MR between 1.3 and 3.3. The range of MR varies with the type of absorber/regenerator, but in general better results are obtained for small MR.

For simulation purposes, validated models are required for modeling the absorber in a liquid desiccant system. Models using lithium chloride have been described by Khan and Martinez (1998), Ahmed et al. (1997) and Kavasogullare et al. (1991). Due to the complexity of the dehumidification process, theoretical modeling relies heavily upon experimental data. Oberg and Goswami (1998b) developed a model for a packed bed liquid desiccant air dehumidifier and regenerator with triethylene glycol as liquid desiccant which was validated satisfactorily by the experimental data. The present study uses a modified version of the mathematical model developed by Oberg and Goswami to compare the experimental results of a packed bed dehumidifier and regenerator using lithium chloride as a desiccant.

例 3[4]

INTRODUCTION

The measure of success of an air conditioning system design is normally assessed by the thermal conditions provided by the system in the occupied zones of a building. Although the thermal condition of the air supply may be finely tuned at the plant to offset the sensible and latent heat loads of the rooms, the thermal condition in the room is ultimately determined by the method of distributing the air into the room. Fanger and Pedersen [1] have shown that the thermal comfort in a room is not only affected by how uniform the air temperature and air velocity are in the occupied zone (the lower part of a room to a height 2m) but also by the turbulence intensity of the air motion and the dominant frequency of the flow fluctuations. There environmental parameters which have profound influence on comfort, are influenced by the method used to diffuse the air into the room. In addition to the supply air velocity and temperature, the size and position of the diffuser in the room have a major influence on the thermal condition in the occupied zone [2].

In air distribution practice, ceilings and walls are very common surfaces which are used for diffusing the air jet so that when this penetrates the occupied zone its velocity would have decayed substantially. Thus the occurrence of draughts is minimized. The region between the ceiling and the occupied zone serves as an entrainment region for the jet which causes a decay of the main jet velocity as a result of the increase in the mass flow rate of the jet.

There are sufficient information and design guides [3, 4] which may be applied for predicting room conditions produced by conventional air distribution methods. However, where non-conventional methods of air supply are employed or where surface protrusions or rough surfaces are used in a wall-jet supply, the design data is scarce. The air distribution system designer has to rely on data obtained from a physical model of the proposed air distribution method. Modifications to these models are then made until the desired conditions are achieved. Apart from being costly and time consuming, physical models are not always possible to construct at full scale. Air distribution studies for the design of atria, theatres, indoor stadiums etc. can only be feasibly conducted with reduced scale models. However, tests carried out in a model should be made with dynamic and thermal similarity if they are to be directly applied to the full scale. This normally requires the equality of the Reynolds number, Re, and the Archimedes number, Ar, [5, 6] which is not possible to achieve in the model concurrently.

The other problem which is often encountered in air distribution design is the interference to the jet from rough surfaces and surface-mounted obstacles such as structural beams, light fittings etc. Previous studies [7, 8] have shown that surface-mounted obstacles cause a faster decay of the jet velocity and when the distance of an obstacle from the air supply is less than a certain value called "the critical distance", a deflection of the jet into the occupied zone takes place. This phenomenon renders the air distribution in the room ineffective in removing the heat load and, as a result, the thermal comfort in the occupied zone deteriorates. Here again there is a scarcity of design data, particularly for non-isothermal air jets.

Air distribution problems, such as those discussed here, are most suitable for numerical solutions which, by their nature, are good design optimization tools. Since most air distribution methods are unique to a particular building a rule of thumb approach is not often a good design practice. For this reason, a mock-up evaluation has so far been the safest design procedure. Therefore, numerical solutions are most suitable for air distribution system design as results can b readily obtained and modifications can be made as required within a short space of time. Because of the complexity of the air flow and heat transfer processes in a room, the numerical solutions to these flow problems use iterative procedures that require large computing time and memory. Therefore, rigorous validation of these solutions is needed before they can be applied to wide ranging air distribution problems.

In this paper a review is given of published work on numerical solutions as applied to room ventilation. The finite volume solution procedure which has been widely used in the past is briefly described and the equations used in the k-ε turbulence model are presented. Numerical solutions are given for two-and three-dimensional flows and, where possible, comparison is made with experimental data. The boundary conditions used in these solutions are also described.

4.3 如何写论文的展开部分(Approach)、结果和讨论(Results and Discussion)

4.3.1 材料和方法部分

对于以实验为主的研究论文，该部分往往位于论文展开部分的前面。

对于实验，描述应尽可能详细。详细的程度应使别的研究者可以重复你的实验，或者对难以重复的实验可评价你的实验。

这一部分经常采用小标题，如：subjects, apparatus, experimental design, chemical synthesis。

在这一部分，应当说明：(1)所用的材料和化学药品的名称；(2)实验条件；(3)实验仪器;(4)实验方法和步骤。

4.3.2 原理和理论模型部分

对于理论分析和数值计算为主的研究论文,该部分往往位于论文展开部分的前面。

一般首先用数学方法描述所讨论的问题,如列出控制方程、边界条件和初始条件。为简化问题并突出问题本质,常需对问题进行合理假设。这部分会引入一些方程、格式、边界条件和初始条件。下面通过一些例子说明其经常采用的表达方式。

例 1[6]

DEVELOPMENT OF MODEL

The model assumes that VOCs are emitted out of a single uniform layer of material slab with VOC-impermeable backing material, and a schematic of the idealized building material slab placed in atmosphere is shown in Figure 1. The governing equation describing the transient diffusion through the slab is

$$\frac{\partial C(x, t)}{\partial t} = D \frac{\partial^2 C(x, t)}{\partial x^2} \quad (1)$$

where $C(x, t)$ is the concentration of the contaminant in the building material slab, t is time, and x is the linear distance. For given contaminant, the mass diffusion coefficient D is assumed to be constant. The initial condition assumes that the compound of interest is uniformly distributed throughout the building material slab, i.e.,

$$C(x, t) = C_0 \quad \text{for } 0 \leqslant x \leqslant L, \ t = 0 \quad (2)$$

where L is the thickness of the slab, and C_0 is the initial contaminant concentration. Since the slab is resting on a VOC-impermeable surface, the boundary condition of the lower surface of the slab is

$$\frac{\partial C(x, t)}{\partial t} = 0, \ t > 0, \ x = 0 \quad (3)$$

A third boundary condition is imposed on the upper surface of the slab (Fig. 1)

$$-D \frac{\partial C(x, t)}{\partial x} = h_m (C_s(t) - C_\infty(t)), \ t > 0, \ x = L \quad (4)$$

where h_m is the convective mass transfer coefficient, m/s; $C_s(t)$ is the concentration of VOC in the air adjacent to the interface; $mg \cdot m^{-3}$; $C_\infty(t)$ is the VOC concentration in atmosphere, $mg \cdot m^{-3}$. It should be mentioned that almost all the physically based models

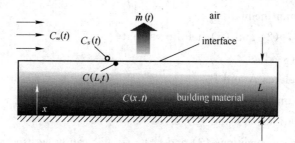

Fig. 1 Schematic shown of a building material slab in atmosphere

in the literature assumed $C_s(t) = C_\infty(t)$, i.e. implied that h_m is infinite, (Dunn, 1987; Clausen et al., 1991; Little et al., 1994). Obviously, the case assumed is a special case of equation (4).

Besides, equilibrium exists between the contaminant concentrations in the surface layer of the slab and the ambient air, or (Little et al., 1994)

$$C(x,t) = KC_s(t), t > 0, x = L. \tag{5}$$

where K is the so-called partition coefficient.

The solutions to equations (1)-(5) derived by us are as follows

$$C(x,t) = KC_\infty(t) + \sum_{m=1}^{\infty} \frac{\sin(\beta_m L)}{\beta_m} \cdot \frac{2(\beta_m^2 + H^2)}{L(\beta_m^2 + H^2) + H} \cdot \cos(\beta_m x) \cdot$$
$$\left[(C_0 - KC_\infty(0)) e^{-D\beta_m^2 t} + \int_0^t e^{-D\beta_m^2 (t-\tau)} \cdot K dC_\infty(\tau) \right] \tag{6}$$

where $H = \frac{h_m}{KD}$, β_m ($m = 1, 2, \cdots$) are the positive roots of

$$\beta_m \cdot \tan(\beta_m L) = H \tag{7}$$

Equation (6) gives the contaminant concentration in the building material slab as a function of distance from the base of the slab, and also of time.

Thus, VOC emission rate per unit area at instant $\dot{m}(t)$ and VOC mass emitted from per unit area of the building material slab before instant $\dot{m}(t)$ can be respectively expressed as follows

$$\dot{m}(t) = -D \cdot \frac{\partial C(x,t)}{\partial x} \bigg|_{x=L} = D \cdot \sum_{m=1}^{\infty} \sin^2(\beta_m L) \cdot \frac{2(\beta_m^2 + H^2)}{L(\beta_m^2 + H^2) + H}$$
$$\left[(C_0 - KC_\infty(0)) e^{-D\beta_m^2 t} + \int_0^t e^{-D\beta_m^2 (t-\tau)} \cdot K dC_\infty(\tau) \right] \tag{8}$$

$$m(t) = -\int_0^t D \cdot \frac{\partial C(x,t)}{\partial x} \bigg|_{x=L} dt = D \int_0^t \sum_{m=1}^{\infty} \sin^2(\beta_m L) \cdot \frac{2(\beta_m^2 + H^2)}{L(\beta_m^2 + H^2) + H} \cdot$$
$$\left[(C_0 - KC_\infty(0)) e^{-D\beta_m^2 t} + \int_0^t e^{-D\beta_m^2 (t-\tau)} \cdot K dC_\infty(\tau) \right] dt \tag{9}$$

下面是引出公式的一些例句、句型：

Plugging these values into…

It yields following inequality：

From Eqs. (1), (2) and (5) it follows that

A may be expressed as

Equation (1) relates A and B.

Then the solution to equation (1) is

Combining equations (1) and (2) gives

Using the boundary conditions (3) and (4), eq. (1) can be written as：

Considering the boundary conditions (3), (4) and (5), the temperature distribution is：

Assuming a relationship between A and B of the form

Expression (1) is applicable only for angles from 0 to θ, where θ satisfies the condition…

Assuming steady-state conditions, governing equations are:…

这里需要注意的是，对公式中出现的符号有两种解释方式，其一是在公式下，用 where 引出解释，其二是在论文中（一般在引言前）用符号表（Nomenclature，Notation，Symbols）说明。一般当符号比较多时采用后者。需注意的是一旦采用后者，公式中出现的符号可不再解释，以避免重复。举例如下：

例 2[2]

NOMENCLATURE

a_t	specific surface area of packing (m^2/m^3)
a_w	wetted surface area of packing (m^2/m^3)
c_p	specific heat (kJ/(kg·℃))
D	diffusivity (m^2/s)
D_p	nominal size of packing (m)
F_G	gas phase mass transfer coefficient (kmol/(m^2·s))
F_L	liquid phase mass transfer coefficient (kmol/(m^2·s))
G	superficial air (gas) flow rate (kg/(m^2·s))
g	acceleration of gravity (m/s^2)
h_G	gas side heat transfer coefficient (kJ/(m^2·s))
k_G	gas phase mass transfer coefficient (kmol/(m^2·s·Pa))
k_L	liquid phase mass transfer coefficient (m/s)
L	superficial desiccant flow rate (kg/(m^2·s))
$LiCl$	lithium chloride
M	molar mass (kg/kmol)
m	flow rate (g/s) or (kg/s)
P	total pressure (Pa)
Pr	Prandtl number
p_v	vapor pressure (Pa)
Sc	Schmidt number
T	temperature (℃)
X	desiccant concentration ($kg_{LiCl}/kg_{solution}$)
x	desiccant mole fraction ($kmol_{LiCl}/kmol_{solution}$)
x_{SM}	logarithmic mean solvent mole fraction difference between the bulk liquid and interface values ($kmol_{LiCl}/kmol_{solution}$)
Y	air humidity ration (kg water/kg dry air)
y	water mole fraction (kmol water/kmol air)

Z	tower height (m)

Greek letters

γ	surface tension (N/m)
ε	effectiveness
λ	latent heat of condensation (kJ/kg)
μ	viscosity (N/m^2)
ρ	density (kg/m^3)

Subscripts

A	air
C	critical
cond	water condensation
equ	equivalent
evap	water evaporation
G	gas phase
IN	inlet
i	interface
L	desiccant or liquid phase
OUT	outlet
o	reference state

例 3[6]

NOTATION

A	emission area of building material (m^2)
C	concentration of compound in building material (mg · m^{-3})
$C_0(x)$	initial concentration of compound in building material (mg · m^{-3})
$C_s(t)$	concentration of compound in the air adjacent to the interface (mg · m^{-3})
$C_\infty(t)$	concentration of compound in atmosphere or in chamber (mg · m^{-3})
D	mass diffusion coefficient for compound in building material (m^2 · s^{-1})
h_m	convective mass transfer coefficient (m · s^{-1})
K	partition coefficient between building material and air (dimensionless)
L	thickness of building material slab (m)
M	total mass per unit area of VOC emitted from building material (mg · m^{-2})
$m(t)$	total mass per unit area of VOC emitted from building material before time t (mg · m^{-2})
$\dot{m}(t)$	emission rate per unit area of VOC from building material at time t (mg · m^{-2} · s^{-1})
PB	particle board

Q	volumetric air flow rate through chamber ($m^3 \cdot s^{-1}$)
t	time (s)
t_c	the critical time (h)
TVOC	total volatile organic compound
V	volume of air in chamber (m^3)
VOC	volatile organic compound
x	linear distance (m)

4.3.3 结果和讨论部分

研究的发现和结果根据其性质不同，有不同的英文表述：Results，experimental results，experimental observations，computer results，numerical results，solutions 以及 results of analysis。研究结果往往用图表表示。

下面的例子说明如何在文中引出图表：

(1) The size and shape of the asci, ascospores and conidia from all specimens examined (Table 1) were essentially identical.

(2) The calculated 37GHz (horizontal polarization) brightness temperatures for dry snow over frozen grounds (Fig. 8) show the effect of the changing thickness of the depth of the hoar layer.

(3) Figure 2 shows the relationship between A and B.

(4) Curve A in Figure 4 illustrates equation (1).

(5) Table 1 summarizes property data for altitudes to 10000m.

(6) The principle of the method may be easily followed from Figure 6.

(7) …is sketched in Fig.1.

(8) …is depicted in Fig.2.

描述观测、测试和实验时，常用一般过去时；讨论数据、图和表时，常用一般现在时。

Results 部分一般介绍的是显而易见的结论；Discussion 部分介绍的是并非显而易见但通过分析和推理可得出的结论，可以对结果进行检验、比较、解释和分析等，并说明结果的意义。

下述表达在该部分会经常使用：

it seems	note that	it is noted that	it appears that
this indicates	it suggests that	it shows that	it provides that
it gives that	it presents that	it summarizes that	it illustrates that
it reveals that	it displays that	it demonstrates that	

上述动词在英文中称为 Indicative or Informative Verbs。

例句：

Table 5 shows the most common modes of infection.

Table 5 shows that the most common source of infection is disks brought from home.

Table 5 provides infection-source percentages.

Inspection of this equation shows that…

The study consists of three parts. The first deals with…, and shows that…The second part covers…, and concludes that…. The third part treats with the…, and demonstrates that….

在结果和讨论表达中，As 从句也经常被使用：

As shown in Table 5, home disks are the most frequent sources of infection.

As can be seen in Fig.8, infant mortality is still high in urban areas.

此外，还需注意动词后介词的使用：

As revealed *by* the graph, the defect rate has declined.

As can be seen *from* the data in Tab.1, …

As shown *by* the data in Tab.1, …

As described *on* page 24, …

注意，这里的 As 从句中一般采用被动语态，从句中无主语！

表示计算或理论分析结果与实验结果相符时，经常采用以下表达：

agree well with, be validate with, be in agreement with, compared favorably with

例句：

(1) The analytical results **compared favorably with** the numerical solution.

(2) A closed-form solution, which could be obtained via a theoretical analysis of the melting process, **shows excellent agreement with** the experimental results.

4.3.4 结论部分

结论部分是全文的重要部分，反映了全文的主要发现和结论，表述力求简明。下面是一些例子：

例 1[1]

CONCLUSIONS

The gravimetric method for directly measuring K and D in VF is simple and effective and can be applied to other indoor materials that can be accommodated in a microbalance. For the compounds and concentration ranges studied, K and D do not depend on concentration. This concentration independence should hold at the lower concentrations typically associated with gas and material-phases in the indoor environment, confirming two of the key assumptions on which the previously developed source/sink diffusion models are based (Little et al., 1994; Little and Hodgsom, 1996; Cox et al., 2000b, 2001b). the observed partition and diffusion coefficients for a series of alkane VOCs correlate well with vapor pressure and molecular weight, respectively, providing a convenient means for estimating K and D for other alkane VOCs in this type of VF without resorting to experimental measurements. Individual VOCs behaved independent of one another during binary sorption experiments, suggesting that the diffusion models may be applied to mixtures of VOCs that are either sorbing or desorbing simultaneously. In contrast, experiments with VOCs and water vapor showed that the presence of sorbed water molecules moderately increases the total uptake of VOCs.

The relatively ideal behavior of the VF studied is somewhat surprising since it is not a perfectly homogeneous material (Cox et al., 200a, 2001a). VOCs in pure polymers generally behave ideally if the concentration of VOCs in the material-phase is lower than 1% by weight (Schwope et al., 1989). The VF material studied here contained about 50% (by weight) calcium carbonate, which might be expected to alter the polymer's behavior. These results are encouraging because they suggest that other relatively homogeneous building materials can be characterized in a similar fashion.

例 2[2]

CONCLUSIONS

Reliable sets of data for air dehumidification and desiccant regeneration using lithium chloride were obtained. The influence of the design variables studied on the water condensation rate from the air and evaporation rate from the desiccant can be assumed linear. Therefore, the slope of the curves in Figs.3-14 give a measurement of the impact of the variable on the water condensation and evaporation rates. Design variables found to have the greatest impact on the performance of the dehumidifier are: desiccant concentration (slope=2.7), desiccant temperature (slope=-1.4), air flow rate (slope=-0.9), and air humidity ratio (slope=2.5). Design variables found to have the greatest impact on the performance of the regenerator are: desiccant temperature (slope=5), desiccant concentration (slope=-1.8), air flow rate (slop=0.5). The other variables have a slope equal or lower than 0.3. in this study the mass flow ratio of air to desiccant solution (MR) varied between 0.15 and 0.25 for both the air dehumidification and desiccant regeneration experiments, which is lower than the MR values of 1.3 to 3.3 used in most other studies. The adapted finite difference model shows very good agreement with the experimental findings.

例 3[5]

CONCLUSIONS

Different from the models in the literature, this proposed model considers both the convective mass transfer resistance through the air phase boundary layer and the distribution of initial concentration in building materials. For given material properties (diffusion coefficient, partition coefficient) and conditions (and the initial concentration and the thickness of the material), it can precisely predict the emissions of VOCs of indoor material for whole process.

The model is validated with experimental data from the literature. By comparing the results of the proposed model with the results derived from the other two models in the literature, it is found that at the early stage, the proposed modeling results agree with the experimental data better while after the initial period, their results are almost the same.

The analysis based upon the proposed model shows that the influence of initial distribution on emission characteristics decreases with increasing time. When time is less than t_c, the influence is strong and should be considered. After that period, such influence can be neglected. And t_c is mainly influenced by L and D. For the long-term emission prediction cases that the material thickness is relatively small and the diffusion coefficient is relatively large, it would be justified to neglect the effect of initial VOC concentration distribution on emission characteristics.

The proposed model together with the analysis is helpful to study the VOC emission characteristics and to improve the measurement precision for VOC emission of building materials.

4.3.5 致谢部分

一般说来，结论部分之后为致谢部分，以说明研究工作得到的基金资助来源，并表示感谢；对研究工作中所得到的其他帮助，在此也应一并感谢。

例 1[1]

Acknowledgements

Financial support for this research was provided by the National Science Foundation (NSF) through a NSF CAREER Award (Grant No. 9624488). We thank Hodgson for his insightful review of the manuscript.

例 2[7]

ACKNOWLEDGEMENTS

This work was supported by National Nature Science Foundation of China (The project grant no. is 50276033) and by Tsinghua University (The project grant no. is 200007005).

5 论文检查和修改

论文完成后，要仔细修改几遍。鲁迅说过，他每次写好文章后，都要改上几遍。大文豪尚且如此，我们对文稿更应仔细修改。

文章写好后，首先应检查题目是否恰当、简明。题目是读者对文章的第一关注点，表达需字斟句酌。

其次，对论文摘要需仔细推敲：是否有遗漏或重复？文字是否准确、精炼？这种推敲一般在文章初稿完成之后进行。

然后对论文的每一部分都要仔细检查、推敲。首先从逻辑、文章结构、内容正确性、表达准确性方面，然后从文字、修辞方面进行仔细检查。总的原则是尽量简明准确，能用一句话说明的决不用一段话，能用一个词说明的决不用一句话。

需要说明的是，文中的数据要注意有效数字，不是位数越多越好。

虽然在文章写作之初，就应注意拟投期刊的投稿要求，因为不同期刊，格式要求可能不一样(包括对参考文献(References)的格式也要注意)。一些易混淆的字母需特别注意：例如 1 (one) and l (ell)，0 (zero) and O (oh)，x (ex) and × (times sign)。

同时，别忘了对图、表的把关，图和表的内容应相对独立、一目了然。其绘制应清晰、精美，数据应准确，坐标轴别忘了标出名称、数值和单位。一般图表不要用灰色背景。很多时候，读者在快速浏览论文时，往往只看题目、摘要、图表和结论，因此，图表在科技论文中举足轻重。

6　根据编辑和评审者的意见修改论文

论文投出并经评审后，编辑会将编辑和审稿人的意见返回。编辑的意见不外乎四种：接受发表、修改后发表、修改后再审和退稿。审稿人的意见会比较具体，指出你文章中的一些问题，你必须逐一仔细回答。回答要实事求是，不能敷衍了事，更不能有不实之词。而且，对审稿人的意见心理上不应首先拒绝，而应充分站在审稿人和读者的立场上想问题。同时对审稿人提出的英语表达上需改正和提高的地方要彻底弄明白，并记下来，以免以后再犯同样或类似的错误。

7　如何提高表达水平

要写好某一领域的英文论文，平时应多读一些该领域一流国际学术期刊的相关文章，读时要做有心人，留心文章结构、文字表达、图、表等，对可能有用的表达应作笔记，以备将来写论文之用。熟读唐诗 300 首，不会吟诗也会吟。读好文章，自然见贤思齐。

以下是笔者从英文学术期刊中记录以备用的几段笔记，仅供参考，虽挂一漏万，但意在表明方法。

引出公式时的常用表达：

(1) For…, we **may define** a mean thickness t **by**

　　$t = \cdots (1)$

　　where…

(2) A **is given** simply **by**

(3) **A was found to be related to B by** the relation：

(4) We have … or We obtain … or We write …

(5) **whence** (＝from which) …

(6) Equation 5 **is replaced by**

(7) The equation **corresponding** to 5 becomes…

(8) Hence…

(9) We readily see from Fig.4 that the x and y **related by**

(10) Equation (5) to (7) may then be written

(11) We obtain **an expression for**…

(12) A may be readily calculated from equation(5) which yields

(13) The equation for…may be simplified by writing…

(14) The expression for the…is…

(15)（微分方程）

　　　　subject(s) to the following boundary condition，

(16) By standard methods，we find the Laplace transform of …

(17) In the present instance we retain dominant terms only and get⋯.

一些常用的表达方式：

(1) Comparison and contrast[8]

by contrast⋯: One dictionary is very good; the other, **by contrast**, is very bad.

on the contrary⋯: Many people think that ours is a good university. **On the contrary**, it is very bad.

on the one hand⋯**on the other hand**⋯: **On the one hand** he is very clever, but **on the other hand** he is vary lazy.

If⋯(**then**)⋯: **If** Jack is clever, **then** John is absolutely brilliant.

Compared with⋯, **in comparison with**⋯: **Compared with** Mike, Peter is extremely brilliant.

(2) Similarity and difference[8]

like⋯: Eire, **like** Spain, is overwhelmingly Catholic.

as with⋯: **As with** all the other problems, I will be dealing with this one very cursorily.

similar to⋯**in** (**respect of**) **etc.**: The two problems **are** very **similar to** each other **in** outlook and temperament.

Punks **are similar to** skinheads **in** their having no hierarchy.

Hong Kong is fairly similar to Taiwan in respect of its economic life.

Different (from, to)⋯**in (respect of) etc.**: Norway and Greese **are different from** each other **in respect of** climate, customs, and economic life.

One (point of) similarity between X and Y⋯ **is that**⋯: One point of similarity between Eire and Spain is that both are overwhelmingly Catholic.

point of similarity between X and Y is⋯

point of resemblance between X and Y is ⋯

point of correspondence between X and Y is⋯

point of divergence of X from Y is⋯

resemble, correspond to, be comparable to, compare (un)favourably with

be analogous to, bear a resemblance to, bear little (or no) resemblance to

(3) Cause and effect

当你对一些现象和结果产生的原因不能确定时，可用如下表述：

It would seem/appear that⋯

It is quite possible that⋯

It is likely that⋯

It is said that⋯

Presumably⋯

There is evidence to suggest that⋯

X may be responsible for Y.

Recent findings suggest that⋯

There appears to be a (statistical correlation) between X and Y.

当你对一些现象和结果产生的原因可以确定时，可用如下表述：

because

as

Since

for

owing to

due to

on account of

because of

therefore

for this reason

hence

consequently

accordingly

as a result (of this)

in view of this

(4) 表示变化趋势或拟合情况

The value of Gn increased as Re increased.

Or：The value of Gn increases with increasing Re.

A logarithmic presentation of the experimental results indicates linear trends for the data distributions.

Linear regressions of the logarithmic values were fitted for the different parts of the data distributions in order to obtain the correlation equations.

(5) 说明研究切入用语

In order to ascertain…

In order to gain insight…

(6) 说明坐标轴情况

The ordinate variable on the right is…. The abscissa variable is ….

(7) 引言中常用语

… has become a subject of current interest.

Much of the earlier work on… was devoted to finding suitable materials.

Many present-day investigations also address this aspect.

一些常用的同义词或反义词：

(1) ensured, obtained

(2) rig, apparatus

(3) agree well with, be validate with, be in agreement with, compared favorably with

(4) however, nevertheless, but, whereas,

Nevertheless, the dilution of average pollution concentrations does not completely eliminates the contaminants and may not solve the problem for local areas of high pollutant emissions or poor air circulation.

However, these techniques transfer the contaminants to another phase rather than eliminating them, and additional disposal or handling steps are subsequently required.

(5) suitable, proper

(6) suitable to, pertaining to

(7) a vast array of, a wide variety of, a broad range of, miscellaneous

A vast array of potential sources of air pollution

A wide variety of contaminants

A broad range of contaminants

Miscellaneous contaminants

(8) destroy, destruct

(9) review, overview

(10) inhibit, promote

(11) Consequently, as a subsequence

(12) In practice, in reality

(13) The previously described model, the aforementioned model

(14) admit, acknowledge

(15) distinctive features, salient features

(16) became increasingly attractive, is of growing interest, is receiving increasing attention in the literature

(17) agree well with that…, conform with that…, … has confirmed the experimental results, accord excellently with that…, be validated with …

(18) because of…, due to…, owing to, resulting from…, because…, the reason …is

(19) drawback, shortcoming, demerit, deficit, flaw

(20) consists of, be composed of, comprise, include, be made up, constitute

Example: These projects constitute the main part.

(21) deal with, treat with, do, undertake, tackle with

Example: be tackled with …method.

(22) explain…, account for …

(23) show, illustrate, describe, demonstrate, indicate

Example: The model results indicate that…

(24) region, range, domain

(25) It is claimed(contended, obvious, seen, reasonable, clear) that….

(26) only, all but

(27) in a clockwise direction, in a counterclockwise direction

(28) in spite of…, despite…, regardless of…, even if(though)…

(29) for the purpose of, for, in order to, in order that, in the interests of

(30) It yields (concludes, is seen from) that…

(31) The objectives of this paper are twofold: first, a computational methodology is presented for predicting the …; second, parametric studies are performed to assess the effects of….

(32) be related to, have sth. to do with, pertinent to

(33) Note that…, It is noted that…

(34) make up for, compensate for, reimburse

(35) need, demand, require, expect

(36) lead to, cause, result in

(37) describe, delineate

(38) and so on, etc., and so forth

(39) Also, a common method of…could be imployed (used).

(40) be in adequate; be lack of; paucity(opposite to glut), scarcity(opposite to glut)

    Example: This invariably leads to glut in the market during the peak periods of production and scarcity during off-seasons.

(41) be fulfilled, be accomplished

(42) for instance, for example, e.g.

(43) Finally, Eventually, Ultimately

(44) as far…is concerned, considering that…

(45) for this case, in such an instance, in mode 1, in case 1, for this scenario

(46) be used to do sth., be assigned to do sth., be imployed to do sth.

(47) respective, corresponding

一些常用的连词：

Furthermore, In addition, Moreover, Besides, However, Nevertheless, Nonetheless

At any rate (in any cases, no matter what happens), Even so

Therefore, For this reason, Consequently, As a result, Hence, Thus

On the other hand

On the contrary

    Example: I believe you like your job. On the contrary, I hate it!

In contrast

    Example: In contrast with/to your belief that we shall fail, I know we shall succeed.

Also, otherwise

Otherwise: 1-if not; 2-apart from that

    Example: 1. Do it now. Otherwise, it will be too late.

            2. He still has a bit of his cold, but otherwise all are well.

In conclusion, After all, Indeed, Actually, in fact, as a matter of fact

Apparently, Certainly, Conversely, Obviously, Undoubtedly, (un)fortunately

* First, * Second, * Then, * Next, * Gradually, * Eventually, * Finally, * Ultimately

* usually they are not used after the words.

最后需要说明的是，写好英文科研论文决定于思维水平、科研水平、知识水平和文学水平。写出好文章，是需要长期修练的。但只要对科研感兴趣，希望提高自己的英文写作水平，通过一段时间的努力，就一定能成功。

练习

(1) 请阅读一篇今年发表的本专业学术期刊论文，注意与"2 科研论文的一般格式"介绍的内容比较。

(2) 根据自己的研究，写一个研究论文提纲。

(3) 请阅读 2～4 篇今年发表的本专业学术期刊论文，注意其 Introduction 的写法，并列出要点。

(4) 请在划线处填上合适的介词。

1) As can be seen _____ figure 4, earnings have decreased.

2) As revealed _____ figure 2, the lightweight materials outperformed traditional metals.

3) As described _____ the previous page, there are two common types of abstracts.

4) As stated _____ Appendix B, per in percent or kilometers per hour is a Latin preposition that originally meant through or by.

5) As described _____ the previous unit, passives are common in process descriptions.

6) As can be seen _____ a comparison of the two tables, household income is a more reliable predictor than level of education.

7) As is often the case _____ materials _____ this type, small cracks pose a serious problem.

8) As has been demonstrated _____ many similar experiments, these materials have many advantages.

(5) 请在划线处填上合适的词或词组。

1) The LA community, _____ many others across the States, had to face the inevitable question.

2) _____ Bankok is ugly, _____ Malina is indescribably hideous.

3) _____ so many things in life, it is best to preserve a certain inner distance from the stupidities of those who govern us.

4) The other courses are terrible, this one _____ is fairly interesting.

5) Such phenomena _____ racism have not found many adherents in Denmark, _____ in Germany.

6) _____ the previous essay, this one also begins with a long quation.

7) _____ it used to be fashionable to have long hair and wide trousers, _____ it will soon be fashionable to have short hair and narrow trousers.

8) Many suppose that the news came as a blow to him. _____, he was delighted.

9) _____ America, Canada can sometimes be a dangerous place for tourists.

10) Highways are still toll-free in this country, _____ Italy. (Italy has highway-tolls.)

11) There are those who, _____ my friend, beg to disagree.

12) This book is interesting; the other, _____, is terribly boring.

13) _____ agricultural and technological revolutions have released the developed world from the Malthusian trap, billions in the developing world are caught in it, condemned to poverty.

14) Many consider him hard-working. _____, he is very lazy.

15) _____ other European countries Spain has its own unique traditions.

16) _____ the previous exercise, this one will also be developed to the future tense.

(6) 请各位自今日始，建立自己的专业英语表达"数据库"。

(7) 根据自己的科研情况，选择合适的题目写一篇英文科研论文。

参考文献

[1] Steven S. Cox, Dongye Zhao, John C. Little, Measuring partition and diffusion coefficients for volatile organic compounds in vinyl flooring, Atmospheric Environment, 35 (2001), PP. 3823-3830

[2] Nelson Fumo, D.Y.Goswami, Study of an aqueous lithum chloride desiccant system: air dehumidification and desiccant regeneration, Solar Energy, Vol. 72, No. 4, pp. 351-361

[3] G. Garrillho da Graca, Q. Chen et al., Simulation of wind-driven ventilative cooling systems for an apartment building in Beijing and Shanghai, Energy and Buildings, 34 (2002), PP. 1-11

[4] H. B. Awbi, Application of computational fluid dynamics in room ventilation, Building and Environment, Vol. 24, No. 1, pp. 73-74, 1989

[5] Ying Xu, Yinping Zhang, An improved mass transfer based model for analyzing VOC emissions from building materials, *Atmospheric Environment*, Vol. 37, No. 18, June, 2003, pp. 2497-2505

[6] Ying Xu, Yinping Zhang, A general model for analyzing VOC emission characteristics from building materials and its application, *Atmospheric Environment*, Vol. 38, No. 1, 2004, pp. 113-119

[7] Yinping Zhang, Rui Yang, Rongyi Zhao, A model for analyzing the performance of photocatalytic air cleaner in removing volatile organic compounds, *Atmospheric Environment*, Vol. 37, No. 24, 2003, pp. 3395-3399

[8] Robin Macpherson 编著. English for Writers and Translators. 北京：清华大学出版社，2003

练习参考答案

LESSON 1　THERMODYNAMICS AND REFRIGERATION CYCLES
请翻译下列句子。

1) 功是指通过存在压差(任一种力)的系统边界传递能量的作用过程,总是指向低压。如果系统中产生的总效果能被简化为一个重物的提升,那么只有功通过了边界。当能量从系统中移出时,功是正的。

2) 一个过程是通过给定初始和最终的平衡状态、路径(如果可以确认)和过程中通过系统边界发生的相互作用来描述。

3) 不可逆概念提供了对循环运行更多的深入理解,例如:在两个固定温度水平间,按给定制冷负荷运行的制冷循环的不可逆性越大,循环运行所需的功也越大。不可逆性包括管道和换热器中的压降、不同温度流体间的传热和机械摩擦。减小一个循环中的总不可逆性则改进循环的性能。在无不可逆性的极限条件下,循环将达到它最大的理想效率。

4) 方程(11)一般应用于流进质量等于流出质量、没有做功并且可以忽略动能和势能的流动系统。

5) 在循环中,动力循环产生的功的减少或者由制冷循环所需的功的增加等于周围绝对温度与循环所有过程中的不可逆性之和的乘积。

6) COP定义为循环收益(除热量)与循环运行所需能量的比值:

$$\text{COP} = \frac{\text{有用的制冷效果}}{\text{外界提供净能量}}$$

7) 这种方法适用于手算和相对简单的计算机模型;然而对很多计算机模拟来说,使用制成表格数据所需的存储或输入输出的日常费用使得这种方法不可接受。对于大的热力系统模拟或者复杂的分析,使用基本的热力学关系或实验数据拟和曲线来确定内能、焓和熵可能更有效。

8) 图5表示出了温熵图上的卡诺循环。热量在恒定温度T_R下从被制冷区域被带出。热量在恒定的周围温度T_0下排放。循环通过一个绝热膨胀和一个绝热压缩完成。

LESSON 2　FLUID FLOW
请翻译下列句子。

1) 流体和固体的区别在于它们对剪切力的反应作用。在施加剪切力时,固体只发生有限的变形,而只要有剪切力的作用流体就会连续变形。液体和气体都是流体。虽然液体和气体的分子运动特性有着很大的区别,但是它们主要的力学区别在于可压缩性的程度和液体自由表面(界面)的形成。

2) 这部分考察均匀、定物性和不可压缩的流体,并且介绍在大多数分析中使用的流体动力学的一些考虑。

3) 对于理想流体模型,物体周围(或者管道截面变化)的流型是由位移效应造成的。流线上的阻碍物,诸如流动中的支杆或者管壁上的突起,推动流动光滑的偏离原来的方向。而在阻碍物之后，流动又恢复一致。流体惯性（密度）的影响只出现在压力改变的地方。

4）在平行管壁的管道中，对于稳定的充分发展的层流，剪切应力 τ 随着距中心线的距离 y 而线性变化。

5）在更一般的边界层流动中，随着散流器中壁面层发展或沿支杆或转向叶片表面的壁面层的发展，压力梯度影响剧烈，甚至导致分离。

6）流体经过收缩管道，如流动喷嘴、文丘里管或者孔口流量计，也可以看作是等熵的过程。

LESSON 3 HEAT TRANSFER

（1）请翻译以下句子，弄清以"-ance"和"-vity"结尾的英语单词的不同物理含义。

　　发射度是物理和电子行业的术语。后缀"ance"表示一块实有材料的性质。结尾"ivity"表示和表面几何条件无关的整体材料的特性。因此发射度、反射比、吸收比和透光度指的是实际材料。发射率、反射率、吸收率和透射率指的是视觉上光滑和足够厚不透明材料的特性。

（2）请翻译下列句子。

1）虽然水在可见光区传输能量，但是因为液体对辐射明显不透明，所以水侧的辐射是不重要的。

2）兰贝特辐射功率变化等价于假设从某一表面非垂直方向产生的辐射，就像是来自于和初始表面有相同发射功率（每单位面积）的辐射。

3）把近来的实验与数值结果和已存在的自然对流传热系数准则进行比较，观察到的差别表明对封闭空间（建筑物）内的垂直表面使用（独立的）垂直平板的传热系数要谨慎。

（3）请用英语表达。

　　in Parallel, in series

　　concurrent, countercurrent

　　thermal resistance, logarithmic mean temperature difference

　　steady state, transient state

　　emission, absorption

　　annulus fin, plate fin, pin fin

LESSON 4 MASS TRANSFER

（1）请把下列词语翻成中文，并熟练使用。

　　空气刷，冷却盘管，蒸发冷凝器，冷却塔

（2）请翻译下列句子。

1）类比的主要限制为：表面形状要一样；且写成无量纲形式时，无量纲温度边界条件类似于 B 成分无量纲密度分布边界条件。几个主要因素使得这样的类推不太精确。在有些情况下，努塞尔数是对光滑表面得出的。很多传质问题包含波形的、小滴形的或者粗糙的表面。多数努塞尔数关系式是对定温表面得到的。由于饱和条件的变化和表面干燥的可能性，有时 ρ_{Bi} 在整个表面上是不恒定的。

2）At temperatures below 60℃ where, Equation(3) can still be used if erron in JB as great as 10% are tolerable.

LESSON 5 PSYCHROMETRICS

(1) 请将下列句子翻译成英文。

1) Saturated humidity ratio $W_s(t, p)$ is the humidity ratio of moist air saturated with respect to water (or ice) at the same temperature t and pressure p.

2) Degree of saturation μ is the ratio of the air humidity ratio W to the humidity ratio W_s of saturated moist air at the same temperature and pressure.

3) Humidity ratio is increased from a given initial value W to the value W_s^* corresponding to saturation at the temperature t^*.

4) Enthalpy is increased from a given initial value h to the value h_s^* corresponding to saturation at the temperature t^*.

5) Mass of water added per unit mass of dry air is $(W_s^* - W)$, which adds energy to the moist air of amount $(W_s^* - W)h_w^*$, where h_w^* denotes the specific enthalpy in kJ/kg (water) of the water added at the temperature t^*.

(2) 请将下列句子翻译成中文。

1) 焓湿学涉及湿空气的热力学，并用这些性质来分析湿空气的状态和过程。Hyland 和 Wexler（1983a，1983b）提出了湿空气和水的热力学性质公式。在大多数空调问题中可以用理想气体关系式来代替这些公式。Threlkeld（1970）提出计算−50℃温度范围内在标准大气压下饱和空气的湿度比、焓和比容时误差小于 0.7%。而且这些误差随着压强的降低而降低。

2) 干湿球温度计由两个温度计组成：一个温度计的感温球用完全被浸湿的纱芯包裹。当湿球被置于空气流中，水从纱芯蒸发，最后达到一平衡温度称为湿球温度。这个过程不是定义热力学湿球温度的绝热饱和过程，而是由湿球同时传热和传质的过程。

3) 对于给定温度的湿球线起点在相应的干球线和两个饱和曲线的交界处，并且它们有相同的斜率。

LESSON 6 THERMAL COMFORT

(1) 请指出下列句子中"subject"所表示的含义。

　　受试者

(2) 请用英语表达。

dry-bulb temperature, wet-bulb temperature, black globe temperature, dew-point temperature

LESSON 7 AIR CONTAMINANTS

(1) 解释下列句子中下划线词语的含义，并翻译整个句子。

1) exposure：暴露

　　在大气压下，即使是在很短的时间内，氧气的浓度小于 12% 或者二氧化碳浓度大于 5% 都是危险的。较少偏离正常成分，暴露时间延长也有危险。

2) secondary：间接的，从初始的或最原始的衍生来的
　　primary：原生的，初生的
　　　　细小的微粒通常在化学性质上较复杂，由人为造成，并且是衍生而来的；而粗大的微粒则大多是原生的、自然的，并在化学上是不活泼的。
3) respirable particle size range：可呼吸粒径范围
　　　　虽然中粒径和大粒径的微粒构成质量的绝大部分，但是80％以上的表面积污染由粒径小于1μm的微粒所引起，而1μm就是可呼吸粒径范围的中心，也是最容易在肺部残留的粒径。
4) population：种群
　　　　微生物种群的过量繁殖一般与这些系统（至少是冷却塔）的不充分的预防性维护有关。
5) survival：生存，存活，not limited to：不限于
　　　　很多因素能够促进军团菌的存活，包括但不仅仅限于：暖和的温度、特定的藻类和原形动物种群、与某些水生植物的共生关系等。
6) resuspension：再次悬浮，disintegrate：（使）分解，（使）碎裂
　　　　如因人体活动而再次悬浮，整粒的花粉可能分解为碎片，而这些碎片可以用能够除去0.3～3.0μm范围内微粒高百分比的高效过滤器（效率为70％～95％，最小效率报告值12～14）来有效控制。
7) breathing zones：呼吸区
　　　　最具有代表性的采样是在呼吸区采集的气溶胶浓度范围的采样。
8) recover：重新得到
　　　　因为军团菌需要特殊的营养素供其生长，并且不会产生有抵抗力的孢子，这种细菌难于从空气中再生。

（2）请用英语表达。

smoke，fog，smog，mist，aerosol

optical particle counter，condensation nucleus particle counter，cascade impactor，electrostatic precipitator

threshold limit value（TLV）

LESSON 8　CENTRAL COOLING AND HEATING

（1）请翻译下列句子。

1) 采用较大的水温升和较低的送风温度，可以减小所需的泵与风机能耗，有时能够补偿由于制冰所需的较低温度而引起的能量损失。
2) 冷却塔也可以通过过滤直接在冷冻水回路循环冷凝器的水，通过一个独立的热交换器冷却冷冻水，或者采用制冷设备的热交换器制冷，在过渡季为建筑供冷。
3) 这些换热表面的有效性取决于一侧的流体、水或蒸汽与另一侧的气体之间的温差和水与热气体的流量。
4) 冷冻水系统可以设计成用三通阀到冷却盘管或用平衡阀到盘管定流量。因流量恒定，系统的负荷与干管送回水温差成正比。
5) 此外，加盖的温度计的套管、试液位旋塞、加盖的管道开口的罩口和流量阀门等都应

该安装在对于系统平衡具有重要作用的位置。

(2) 通过阅读文章，请说明 Central Heating and Cooling system 的特点。

　　集中式供热与供冷系统能源输入可以是电、油、燃气、煤、太阳能、地热等，这些能源被集中转换为热水（或蒸汽）和冷水，然后被输送到建筑的各个需要供热和供冷的空间。集中式冷热源系统可以采用多样化的能源，与分散式系统相比其效率较高、维护与劳动力费用较低；但它也需要较大的主机机房空间，输配系统也比较庞大，有时灵活性不及分散式系统。

LESSON 9　DECENTRALIZED COOLING AND HEATING

(1) 请翻译下列句子。

1) 风量减小时，通常可以调高机组的出风温度，以维持最小的通风量。重新调节出风温度是一种限制机组需求量的方法，节省能量。

2) 因为不需要能量从设备间传送空气或冷冻水，能耗可能低于集中式系统，但是，这种优点可能被这种设备的低效所抵消。

3) 在存在烟囱效应的建筑中，穿墙式空调器的使用应限于在有可靠通风和内区与外区之间用隔墙隔开的区域。

(2) 请用英语描述。

unitary air conditioner, split air conditioner, window air conditioner, packaged air conditioner

air cooled condenser, evaporative condenser

centrifugal fan, axial fan, forward curved fan, backward curved fan, airfoil fan

LESSON 10　AIR-COOLING AND DEHUMIDIFYING COIL

(1) 请翻译下列句子。

1) 等边形（交错排列）或矩形（成行排列）中心管间距在 15～75mm 之间，取决于肋片的宽度和其他性能的考虑因素。

2) 有些盘管制造厂家提供可拆卸的水联箱板，或者每根管都有可拆卸的塞子，使盘管能够被清洁，保证盘管在工作时额定性能的延续。

3) 如果盘管负荷不能均匀分配，盘管就该用多于一个热力膨胀阀给各回路供液来重新安排回路和连接（分开吸气也可能有帮助）。

4) 盘管的设计特点（肋片间距、肋管间距、迎风面高度、肋片类型），与盘管上的凝结水量和表面的清洁度决定了将盘管上凝结水吹出的空气流速。

5) 空气冷却和除湿盘管的框架以及凝结水盘和水槽应采用适合于系统及其期望使用寿命的可接受的防腐蚀材料。

(2) 请写出英文。

　　capillary tube, thermostatic expansion valve (TXV), constant pressure expansion valve, solenoid valve; superheat, cooling coil, eliminator plate, drain pan, face velocity, face area, surface tension

LESSON 11　AIR CLEANERS FOR PARTICULATE CONTAMINANTS

（1）请解释下列句子中下划线词语的含义，并翻译整个句子。

1）concentration gradient：浓度梯度

当越来越多的粒子被捕捉，在纤维附件的区域就会形成浓度梯度，进一步通过扩散作用和拦截作用加强过滤。

2）load：捕集了粉尘

当介质是新的和清洁的时候，过滤器的效率一般认为是最高的，当其荷尘时效率就很快降低。

3）obscure：使模糊，遮掩

但是，这些困难都不应遮掩目的，及设备性能测试应尽可能地接近模拟实际使用条件的目的。

4）versus：与…相对

上、下游的空气采样通过采用光学粒子计数器或相似的测量仪器在规定的气流量下来获得与粒径相对的除尘效率。

5）overloading：超载

变风量和定风量系统的控制，防止了在过滤器清洁时发生的不正常的高风量或者可能的风扇电机超载。

6）potential：电压

在电离段，带有6～25kV正电直流电压的细金属丝悬挂在接地板的等距离处。

（2）请用英语表达。

low-efficiency air filter, medium-efficiency air filter, sub-high efficiency air filter, high-efficiency particulate air filter (HEPA filter), ultralow penetration air filter (UPAF), viscous impingement air filter, dry-type air filter, electronic air cleaner, automatically-renewable media roll filter, prefilter, after filter, final filter, main filter, charged media air filter, panel air filter

LESSON 12　UNITARY AIR CONDITIONERS AND UNITARY HEAT PUMPS

（1）请解释下列句子中下划线词语的含义，并翻译整个句子。

1）structural adequacy：结构负载足够度

大容量的设备应该仅在评估了屋顶的结构适合之后，才能在屋顶安装。

2）box in：包围

不要用栅栏、围墙、挑檐或树丛将室外风冷机组围起来，这样做会减小机组的空气流通能力，从而降低效率。

3）heat sink：热汇，heat source：热源

在制冷工况充当热汇的盘管，在制热工况充当热源。

4）hot deck：制热段，cold deck：制冷段

这些机组中的空气通路被设计成使送风经过一个有加热手段的热段，或者经过一个通常装有一个直接膨胀式蒸发盘管的冷段。

5) sweat：结露

标准要求机组能够合理地排出凝结水，而且机壳在凉湿的环境下不结露。

(2) 请用英语表达。

air-source heat pump, water source heat pump, wter loop heat pump, ground source heat pump, unitary air conditioner, split air conditioner, seasonal energy efficiency ratio (SEER), heating seasonal performance factor (HSPF)

英文科技论文写作简介练习答案

练习 4：

1) from/in, 2) by/in, 3) on, 4) in, 5) in, 6) from, 7) with, of, 8) by/in。

练习 5：

1) like（not "similarly to"）, 2) If…then…, 3) As with, 4) by contrast, 5) as, unlike, 6) Like, As with（not "similarly to"）, 7) If, then, 8) On the contrary, 9) Like（not "similarlly to"）, 10) unlike, 11) like（not "similarly to"）, 12) by contrast, 13) If（also：While, Whereas）, 14) On the contrary, 15) Like（not "similary to"）, 16) As with, Like

附录

附录 A 国际相关组织介绍

1. American Society of Heating Refrigerating and Air-conditioning Engineers,简称 ASHRAE

美国供热、制冷与空调工程师学会(ASHRAE)致力于促进供热、通风、空调、制冷及相关人性因素的科学和技术的研究与发展,以满足公众及 ASHRAE 会员对该领域科学和技术不断进步的要求。学会历史上由两个协会的管理者共同创建,即美国供热、通风工程协会(ASHVE)(1954 年改名为美国供热、空调协会(ASHAE))和美国制冷工程协会(ASRE)。这两个协会在 1959 年合并为美国供热、制冷与空调工程师学会(ASHRAE)。

网址:www.ashrae.org

2. Air & Waste Management Association,简称 A&WMA

空气和废弃物管理协会(A&WMA)由加拿大和美国的烟检人员始创于 1907 年,是一个非营利性专业组织。成员包括科学家、工程师、决策者、律师以及政府、企业和大学的相关顾问。协会的宗旨是通过一个中立的信息交流论坛来发展改进环境科学与技术,完善相应的知识和法规,在重大的环境决策上为政府和社会提供帮助。

网址:www.awma.org

3. International Society of Indoor Air Quality and Climate,简称 ISIAQ

国际室内空气品质和气候协会(ISIAQ)始创于 1992 年。在第 5 届室内空气品质和气候国际会议(多伦多,1990 年)后,由 109 位国际科学家和从业者所创。

国际室内空气品质和气候协会是一个国际性的、独立的、多学科的非营利性组织,宗旨是通过以下方面为创造健康、舒适和有利于提高生产率的室内环境提供支持:

1) 推进室内空气品质领域科学和技术的发展,内容涉及室内环境设计、施工、系统运行和维护以及空气品质测试和对人健康的影响;

2) 在室内空气品质和气候方面,通过出版和促进发行,推动国际和相应学科间的科技交流;

3) 组织、主办和赞助室内空气品质和气候方面的各种学术会议;

4) 为室内空气品质和气候的发展制定、改编和修订法规、标准和政策;

5) 与政府和其他部门以及对室内环境感兴趣的社团进行合作。

网址:www.isiaq.org

附录 B 有关国际会议简介

1. ASHRAE 年会

由 ASHRAE 主办，每年两次，分为冬季年会（每年 1 月份在美国举办）和夏季年会（每年 6 月份在美国举办）。冬季年会展览，展出一些暖通空调领域的新产品和新技术，有厂家参展。会后，一些会议论文编入 ASHRAE Transactions 中。详细信息可登录 ASHRAE 网站：www.ashrae.org。

2. A&WMA 年会

由 A&WMA 主办，每年一次。详细信息可登录 A&WMA 网站：www.awma.org。

3. Indoor Air

从 1978 年开始，三年一届的国际室内空气会议汇集了广大室内空气品质和气候的研究人员，致力于理解和解决室内空气品质和气候的相关问题。最新一届会议（Indoor Air 2002）在美国加利福尼亚州的蒙特立举办。第十届国际室内空气品质和气候会议（Indoor Air 2005）将在中国北京举行（9 月 4~9 日）。有关此 2 届会议的情况请登录：

www.indoorair2005.org.cn（第十届国际室内空气品质和气候会议）

www.indoorair2002.org（第九届国际室内空气品质和气候会议）

4. Healthy Building

健康建筑国际会议起始于 1988 年，目的是将健康建筑的研究成果应用到实际中去。第一届健康建筑会议在瑞典首都斯德哥尔摩举办。此后，华盛顿、布达佩斯、米兰、奥斯陆和埃斯波先后举办过此国际会议。最近一次会议于 2003 年在新加坡举行。

会议的目的是将室内空气品质研究的最新知识、最新建筑技术以及产品开发应用到经济而安全的健康居室和工作环境中去。

2003 年会议信息可登录网站：www.hb2003.org。

5. ROOMVENT

从创立开始，ROOMVENT 系列会议一直为从事空气流动的研究人员提供了良好的国际交流机会。会议召集了来自大专院校、国家研究协会以及工业界科学家和工程师来介绍和探讨各自领域的最新成果。

ROOMVENT 会议起始于 1987 年，由 SCANVAC（丹麦、芬兰、冰岛、挪威和瑞典组成斯堪的纳维亚暖通清洁工程协会联盟）在斯德哥尔摩举办。第二届 ROOMVENT 会议于 1990 年在挪威奥斯陆举办，从那以后，ROOMVENT 会议每两年举办一次，先后在奥尔堡（丹麦，1992）、克拉科夫（波兰，1994）、横滨（日本，1996）、斯德哥尔摩（瑞典，1998）、里丁（英国，2000）、哥本哈根（丹麦，2002）和可因布拉（葡萄牙，2004）。详细情况请登录：www.roomvent2004.com。

6. ISHVAC

由清华大学建筑技术科学系发起主办、由中国国家自然科学基金资助的 ISHVAC（国际暖通空调研讨会）始办于 1991 年。以后每四年举办一次，第二、第三次会议分别在北京和深圳举行。最近一次会议即 2003 年第四届国际暖通空调研讨会在北京同方科技大厦举行。会议的目的是聚集 HVAC 相关领域的国内外专家交流思想、理念以及研究成果，这些专家包括大学和研究协会的研究人员、工程师以及工业、政府部门的有关人员。

详细情况可登录：www.ishvac2003.org

附录C 相关领域的一些国际期刊简介

1. ATMOSPHERIC ENVIRONMENT

"大气环境"出版的论文主题涵盖了人与生态系统大气环境相互作用的所有方面。包括科学、管理、经济和政治等多方面的相互关系。"大气环境"的主要目的是为自然因素和人为因素对大气的影响提供科学证明。研究领域包括但不局限于以下一些方面：空气污染的研究及应用，空气品质及其影响，污染物传播和输运、沉积作用，生物-大气的物质交换，地球大气化学，辐射和气候等等。基于实验的新论点、从局部的到全球范围的大气理论和模型都在本刊范围之内。影响因子：2.352。

详细情况请登录：www.elsevier.com/wps/find/journaldescription.cws_home

2. INDOOR AND BUILT ENVIRONMENT

"室内及人工环境"出版下列方面的研究论文：室内和建筑物内的环境品质，以及它们对健康、工作、效率和人体舒适性可能造成的影响。论题范围还包括城市基建、建筑设计以及用于实验研究的材料（包括动物模拟和人造环境效果实验）。影响因子：0.608。

详细情况请登录：www.sagepub.com

3. BUILDING AND ENVIRONMENT

"建筑和环境"出版的文章内容包括建筑研究及其应用的研究成果，内容涉及以下方面：

- 建筑物的环境特性、组成和原材料
- 室内气候设计和性能
- 人体对室内外物理环境的反应
- 环境设计方法和技术，包括计算机模拟
- 建筑设计、规划、政策制定的应用实例研究
- 建筑环境方面的维护和再利用
- 本地传统建筑和住宅的环境参数（包括经济、社会和文化等相关方面）
- 国际技术和本地传统的综合
- 建筑研究和建筑科学的原理和策略
- 建筑科学和技术的历史
- 建筑设计专业的教育情况

影响因子：0.554。详细情况请登录：www.elsevier.com/wps/find/journaldescription.cws__home

4. ENERGY AND BUILDINGS

"能源和建筑"是出版与建筑物能源利用相关文章的国际期刊。目的是介绍建筑节能及提高室内环境品质方面的创新研究成果和实验结果。

涉及的方面有：

- 现有建筑和未来建筑的能源需求和消耗
- 热舒适性和室内空气品质
- 自然通风和机械通风

- 空调建筑内的气流组织
- 太阳能及其他可再生能源在建筑中的应用
- 大空间建筑的能量平衡(工厂、公共建筑等等)
- 住宅、公共建筑及工业建筑的暖通空调和制冷系统
- 建筑热回收系统
- 建筑物和区域供热
- 人工环境的能量守恒
- 节能建筑
- 建筑物理
- 建筑围护结构和能耗
- 室内冷热和照明系统的评价和控制
- 智能建筑
- 建筑设计和机械、照明系统的关系
- 建筑新材料及其对能耗的影响
- 节能建筑的内外部设计

基于模拟得出结果的论文同样受到欢迎,尤其是那些与实验或实际测量有明显关系的将更受欢迎。影响因子：0.500。

详细情况请登录：www.elsevier.com/wps/find/journaldescription.cws_home

5. INDOOR AIR - INTERNATIONAL JOURNAL OF INDOOR AIR QUALITY AND CLIMATE

本刊是非工业建筑室内环境方面独创研究成果报道的平台。出版的论文包括以下方面：室内空气品质对健康的影响,热舒适性,监测和建模,污染源的描述,通风及其他控制技术等。研究结果将为建筑设计师、业主及经营者提供参考信息,以给居住者提供一个健康舒适的环境,也给医学家对怎样处理与室内环境相关的病症提供信息。影响因子：1.516。

详细情况请登录：www.blackwellpublishing.com

6. ENVIRONMENTAL SCIENCE & TECHNOLOGY

"环境科学与技术"是广大环境学科专业人员权威的信息来源之一。致力于出版自然和人工环境方面的科研文章,尤其关注环境中的化学作用,包括自然和人为两个方面,也包括生理现象以及与认识、控制环境相关的数学和计算方法,同样收录描述改善、控制和预防污染的重大技术进展的文章。其涉及内容如下：自然和人工环境的描述、环境的运转过程、环境的评价方法、改善和控制技术、可持续工程和绿色化学、政策以及文章评论。影响因子：3.123。

详细情况请登录：pubs.acs.org/journals/esthag

7. ENERGY JOURNAL

"Energy Journal"是 IAEE 的官方季刊。它创办于 1980 年,目的是促进能源及相关方面知识的进步和传播。编者们一直致力于出版一系列融合理论、实际经验和政策的文章。每期季刊共 150 页,包含与能源相关的研究文章、短讯及综述文章。包括的范围和主题有：

- 能源和环境问题
- 石油
- 电力行情
- 发展中国家和能源
- 天然气
- 汽油需求分析
- OPEC 和石油市场
- 可再生能源
- 方针政策
- 煤炭问题
- 分布的形成
- 经济模型
- 运输燃料的选择
- 能量的效率
- 经济学的规章制定
- 能源税收
- 市场力研究
- 传统能源的取代
- 核能问题
- 运输
- 散发物的交换（SO_2，CO_2）
- CO_2 散发量的降低

影响因子：0.814。详细情况请登录：www.iaee.org/en/publications/journal.aspx

8. ENERGY

本期刊是多学科综合的焦点，集中于能源相关项目的发展、评估和管理。主要包括：节能系统的输入输出分析，能量维持的测量及其实现，能源系统管理评价，环境影响评价，强调经济性的方案选择等。影响因子：0.564。

详细情况请登录：authors.elsevier.com

9. HVAC&R RESEARCH

"International Journal of Heating, Ventilating, Air-Conditioning and Refrigerating Research"是一份包含国际知名专家深刻见解的季刊。其文章的评定基于它对R&D的价值以及是否将研究描述得足够详细以对其他研究人员的工作产生帮助的程度。本刊始于1995年，出版文章包含全球性的主题，如：影响室内空气品质的散发源，模拟制冷剂和润滑油混合物的热力学方法，以及能量维持策略等。期刊影响因子：0.630。

详细情况请登录：www.ashrae.org/template/EducationLinkLanding;/category

10. JOURNAL OF THE AIR & WASTE MANAGEMENT ASSOCIATION

"空气和废物管理协会期刊"是全球最老的环境技术期刊之一。最早以"Air Repair"的名字发行于1951年。本刊旨在为空气污染控制和废物处理相关专业人员提供服务，提供及时而可靠的信息。影响因子：1.496。

详细情况请登录：www.awma.org/journal

11. ASHRAE JOURNAL

"ASHRAE Journal"是美国供热、制冷与空调工程师学会出版的期刊，主要面向应用研究，发表较宽范围内 HVAC&R 技术性研究论文。它所包含的内容从描述基本特性到综述新兴技术，包括了所有 HVAC&R 应用的范围。考虑到其权威性、平衡性和实用性的编辑内容，它是一本美国供热、制冷与空调工程师学会会员都会收到的月刊。其读者包括顾问工程师、机械承包人和建筑设计公司雇佣的工程师、建筑师、操作工程师和负责 HVAC&R 服务的车间工程师以及由 OEMs 与其他一些设计研究和开发的机构雇佣的设计工程师。影响因子：0.195。

详细情况请登录：www.ashrae.org/template/JournalLanding

附录 D EI 和 SCI 检索的简易教程

1 什么是 EI 和 SCI

工程索引数据库(Engineering Index)简称 EI。EI 数据库由美国工程信息公司建立，该数据库从大约 40 个国家 26 种语言 4500 种文献源中(包括期刊、科技报告、会议录和专著等)，精选出高质量的科技文章，予以报道，其专业覆盖面十分广泛，包括能源、环境科学、地理学、生物学、电子学、自动控制、原子技术、航天航空技术、计算机技术、工业机器人以及土木工程技术等，是很有价值的数据库。

科学引文索引数据库(Science Citation Index)简称 SCI。SCI 数据库是美国费城科学情报所建立的引文数据库之一。该数据库精心挑选了 3800 种有代表性的权威科技期刊作为数据源，并声称这些数据源包括了世界上 90% 以上的重要的科技文献，所以被它收录的论文具有较高的质量，代表了当时有关领域的先进水平。该数据库约有 1600 万条记录。

2 如何利用网络以及 EI 和 SCI 检索系统进行文献检索和获取有用信息

通过一些网站，可进入 EI 和 SCI 检索系统。下面以通过清华大学校园网站为例，说明 EI 和 SCI 检索系统的使用。

2.1 进入"电子资源"库

进入清华大学图书馆主页 http://lib.tsinghua.edu.cn，在"电子资源"中找到"数据库"一栏(如图 1)。点击"数据库"，进入电子资源和数据库导航，在"常用数据库"中可以找到各种数据库，其中包括 EI 和 ISI Web of Science (SCI, SSCI, AHCI)，见图 2。

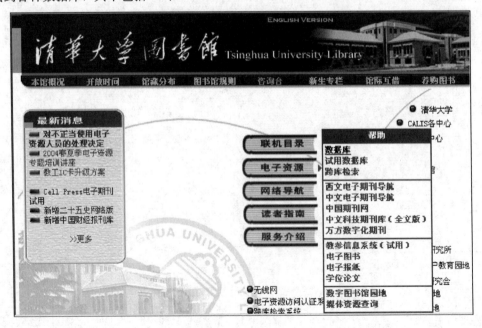

图 1 清华大学图书馆主页

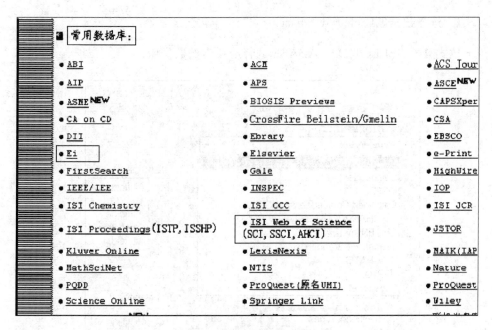

图 2　电子资源和数据库导航

2.2　EI 检索

点击 EI，进入 EI 清华大学图书馆镜像服务页，点击 Engineering Village 2 链接进入 EI 快速检索页面，见图 3。

图 3　EI 快速检索

检索说明如下：（以下内容摘录自清华大学图书馆网页）

2.2.1 选择数据库(Select database)

用下拉式菜单 SELECT DATABASE 选择要检索的数据库。

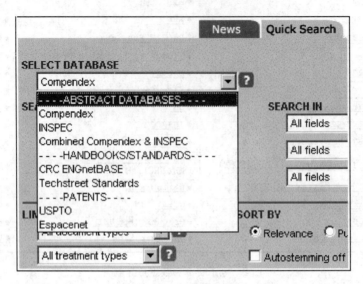

图 4 选择数据库

(1) Compendex 数据库(Compendex)

Compendex 数据库是目前全球最全面的工程检索二次文献数据库，包含选自 5000 多种工程类期刊、会议论文集和技术报告的超过 7000000 篇论文的参考文献和摘要。

网上可以检索到 1970 年至今的文献。

(2) INSPEC 数据库(INSPEC)

INSPEC 为一流的文献数据库，通过它可以访问世界上关于电气工程、电子工程、物理、控制工程、信息技术、通信、计算机和计算等方面的科技文献。

在网上可以检索到 1969 年至今的文献记录。此数据库每周更新数据。

INSPEC 数据库由 the Institution of Electrical Engineers 编制。

(3) Compendex 和 INSPEC 联合检索(Combined Compendex & INSPEC)

如果用户所在单位购买了 INSPEC 数据库，选择 Combined Compendex & INSPEC，可以联合检索 Compendex 数据库和 INSPEC 数据库中所有的科学、应用科学和工程技术学科的相关题目。

Combined Compendex & INSPEC 数据库每周更新数据。

(4) CRC ENGnetBASE

如果用户所在的机构购买了 ENGnetBASE 数据库，则用户就可以访问由 CRC 出版的世界上一流的工程手册。ENGnetBASE 数据库包含可从网上检索到的超过 145 部此类手册，而且一旦有新书出版或更新，将会更多。如果需要 ENGnetBASE 手册的目录，可访问网站：www.engnetbase.com。

(5) Techstreet 标准(Techstreet Standards)

Techstreet 是世界上最大的工业标准集之一，收集了世界上 350 个主要的标准制定机构所制定的工业标准及规范。关于 Techstreet 更详细的信息，可访问其网站：www.techstreet.com。

(6) USPTO 专利(USPTO Patents)

选择 USPTO Patents 可以访问美国专利和商标局（The United States Patent and Trademark Office)的全文专利数据库。在此可以查找到 1790 年以来的专利全文，此数据库的内容也是每周更新一次。

1790~1975 年间的专利只能通过专利号或目前的美国专利分类码检索得到。关于此数据库的更详细信息可访问其网站：www.uspto.gov/patft/index.html。

要查找用户在 Compendex、INSPEC、esp@cenet 或 Scirus 数据库中检索到的有关流程、工艺和产品的专利，只需把所要查寻的关键词输入到 Engineering Village 2 中 USPTO 检索栏中，此关键词就被送到 USPTO 的站点，用户就可以浏览与所检索主题相匹配的专利的详细背景信息。

(7) esp@cenet

通过 esp@cenet 可以查找在欧洲各国家专利局及欧洲专利局（EPO）、世界知识产权组织（WIPO)和日本所登记的专利，关于此数据库的更详细信息可以访问其网站：ep.espacenet.com。

(8) Scirus

Scirus 是迄今为止在因特网上最全面的科技专用搜索引擎。采用最新的搜索引擎技术，由此科研人员、学生及任何人可以准确地查找科技信息、确定大学网址、简单快速查找所需的文献或报告。

2.2.2　检索基础(Search basics)

(1) 检索词检索

将要检索的词或短语输入一个或几个 SEARCH FOR 文本框中，用户也可从文本框右边的 SEARCH IN 下拉式菜单中选定字段进行检索。

此界面有三个检索框，允许用户将输入不同检索框中的词用布尔运算符 AND、OR 和 NOT 连接起来，进行联合检索。

为了放宽检索条件或检索有不同拼写方法的同一个词，可以用布尔运算符 OR 将词连接起来(可得到包含这些词中任何一个的检索结果)。

例如：rapid transit OR light rail OR subways. seatbelts OR seat belts

为了缩小检索范围，可以用布尔运算符 AND 将词连接起来(得到只有包含所有这些词的检索结果)。

例如：prosthetics AND biocontrol。

可以用布尔运算符 NOT 删除包含某些词的检索结果。

例如：为了检索作为建筑物一部分的 windows(窗口)，而不是 Microsoft windows(视窗操作系统)，可输入：windows NOT Microsoft。

如果三个文本框中均有输入，快速检索(Quick Search)总是先合并检索前两个文本框中的词，然后再检索第三个文本框中的词。

a AND b OR c 检索的顺序为 (a AND b) OR c

a OR b AND c 检索的顺序为 (a OR b) AND c
a OR b NOT c 检索的顺序为 (a OR b) NOT c
举例：如图 5 所示。

图 5　检索基础

本例中检索结果包含 International Space Station 或者 Mir，但均须包含 gravitational effects。如果用户需要在一次检索中使用更多的词，请用高级检索(Expert Search)，或者点击上部导航条(top navigation)中的检索历史(Search History)按钮，启用合并以前的检索(Combining Previous Searches)功能。

(2) 说明

• 大小写(Case sensitivity)

Engineering Village 2 的界面不区分大小写，所输入的单词可以是大写也可以是小写。

• 排序(Sorting)

Compendex、INSPEC 和 Combined Compendex & INSPEC 数据库的检索结果可以按相关性(relevance)或按出版时间进行排序。默认的排序为相关性排序。

• 相关性(Relevance)

相关性排序基于以下准则：

这些词是作为一个精确的短语检索到的还是该短语中的词在一条记录中被分别检索到的；

如果这些词是分别检索到的，被检索到的词越接近，该条排列越靠前；

词或短语在检索到的记录中出现的次数；

词在文档中的位置(在文档开始字段中发现的则排在前，靠近末尾的则排在后)；

此词是否是在重要的字段中检索到的，例如标题字段。

• 出版年份(Publication Year)：

按记录的出版时间进行排序，新近出版的文献排在前面。

• 复位(Reset)

当用户需要在检索过程中开始一次新的检索，请点击复位(reset)按钮，清除前面的检索结果。点击复位(reset)按钮可确保前面的检索结果不影响新开始的检索，并且将所有的选项复位到默认值。

2.3 SCI 检索

2.3.1 进入 SCI 检索界面

在电子资源和数据库导航(见图 2)中点击 ISI Web of Science (SCI，SSCI，AHCI)，进入 Web of Science 镜像页，点击 Web of Science(SCI、SSCI、A&HCI、CCR、IC)链接进入 ISI Web of Science 主页。

在下拉式菜单中选择数据库 ISI Web of Science，点击右侧按钮 GO(见图 6)。在 ISI Web of Science 页面中点击"FULL SEARCH"按钮，在 FULL SEARCH 页面中可以选择数据库 SCI 和搜索时间范围(见图 7)，点击下面的"GENERAL SEARCH"按钮开始检索。

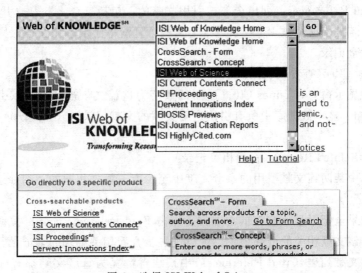

图 6　选择 ISI Web of Science

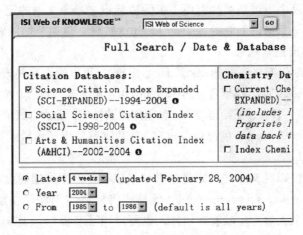

图 7　选择数据库和年份

在相应检索框中输入检索词(支持逻辑算符、位置算符和通配符)，选择限制条件后就可以开始检索了。

2.3.2 检索结果的利用(以下内容摘录自清华大学图书馆网页)

(1) 浏览检索结果

实施 Easy Search 和 General Search 检索后,首先分页(每页 10 条)显示检索结果的简单记录,简单记录包括文献的前三位著者、文献标题、出版物名称(刊名)、卷、期、起止页码和出版时间。点击文献标题,可以看到该条结果的全记录(full record),包含记录的全部字段。在全记录显示窗口的左上方,Previous、Next 和 Summary 三个按钮的作用分别为转到上一条全记录、下一条全记录和分页列出简单记录。利用翻页按钮迅速跳到希望查看的页码。

(2) Cited References(查看引用文献,即参考文献)

在全记录显示窗口,Cited References 后面的数字为这篇文献所引用文献的数目。

点击 Cited References,列出全部被引用文献的,包括著者、发表被引文献的出版物、卷、页和年。若被引文献被 ISI 收录进某个文档,且清华已定购该文档,则可以通过点击查看被引文献全记录。

(3) Times Cited(查看被引用次数)

在全记录显示窗口,Times Cited 后面的数字为这篇文献被其他文献引用的次数。点击 Times Cited,显示数据库中所有引用这篇文献的文献记录(Citing Articles)。显示格式为简单记录。

(4) Find Related Records(查找相关记录)

若数据库中某两篇文献引用的参考文献中至少有一篇是相同的,则称这两篇文献为相关记录。

在检索结果全记录显示窗口的右上方,点击 Find Related Records 按钮,列出在已定购文档中,与该文献相关的全部记录。可以进一步查看相关记录的全记录。

在被引用文献列表(Cited References)中,缺省情况为每篇文献都有标记,此时点击窗口右上方的 Find Related Records,查找的相关记录同(1);若去除一篇或数篇文献的标记,查到的相关记录中不包括那些引用了去除掉标记之参考文献的记录。

点击相关记录显示窗口上方的 Search Results 按钮,返回原始检索结果显示窗口。

(5) Export(输出检索结果)

在显示标记结果的窗口(点击 Marked List 后出现),可以输出这些记录。输出的格式可以选择,缺省格式为简单记录,检索者可选择增加引用参考文献、地址、文摘等字段。输出多记录时可以按时间、第一著者、原始出版物或被引用次数排序。输出方式可选择以下之一:

Foramt for Print——以简单记录格式在 WEB 浏览器中显示检索结果,借助浏览器的打印功能打印。

Save to File——以纯文本的格式将检索结果保存在检索终端硬盘或软盘上。

Export to Reference software——以 cgi 格式将检索结果保存在硬盘或软盘上,保存的文件可以输入到个人参考文献管理软件,也可以用写字板(wordpad)等软件打开。

Email——以电子邮件的形式将检索结果送出。

下面通过举例来具体说明:

例如要查找 2003 年关于 Indoor Air Quality 或 IAQ 的 SCI 文献中与通风(ventilation)相关的所有文献。

首先在引用数据库中选择 SCI-EXPANDED，在时间范围中选择 2003 年，然后点击 GENERAL SEARCH，如图 8。

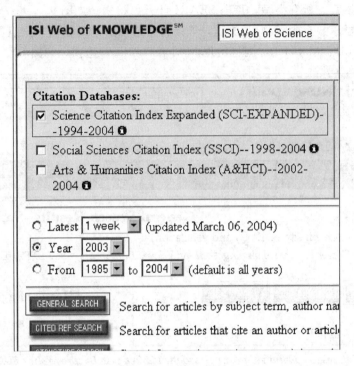

图 8　检索范围选择

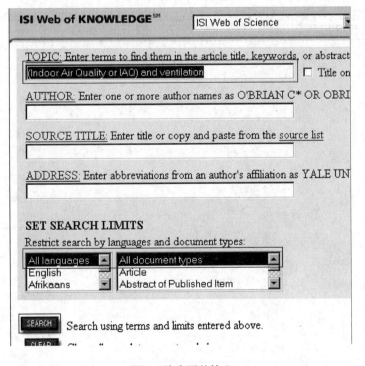

图 9　检索词的输入

在检索页面的检索框中输入：(Indoor Air Quality or IAQ) and ventilation，表示关于 Indoor Air Quality 或 IAQ 的 SCI 文献中所有与通风(ventilation)相关的文献。在检索范围中选择所有语言、所有类型，点击 SEARCH 开始检索。如图 9。

检索结果显示共有 48 篇此类文献，如果将结果想要全部输出，则点击 MARK ALL，然后点击 SUBMIT MARKS，则将出现 MARKED LIST 选项，如图 10。

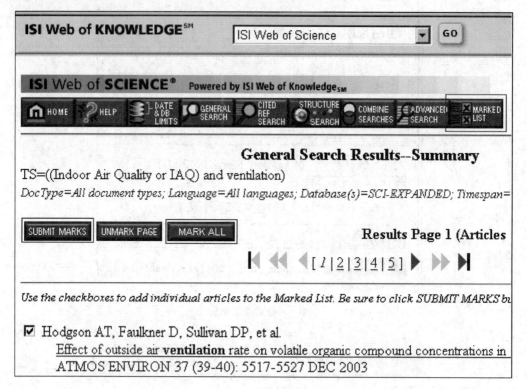

图 10　结果的标记